AF564999

Soil Resources Inventory Management

About the Authors

Dr. V. Arunkumar is presently working as Assistant Professor in the Department of Soil Science & Agricultural Chemistry, Agricultural College & Research Institute, Vazhavachanur, Tamil Nadu Agriculural University. He studied B.Sc. (Ag) degree from Annamalai University and M.Sc. (Ag) and Ph.D from Tamil Nadu Agricultural University, Coimbatore. He was awarded with Senior Research Fellowship in the project "Remote sensing for Agricultural Applications" funded by Government of Tamil Nadu for his Doctoral degree programme. He has more than 15 years of research experience in Natural resource Management. He has organized a National workshop on "Role of Potassium Nutrition to Sustain Crop Production in Tamil Nadu" funded by International Plant Nutrition Institute. He has operated one International project funded by International Plant Nutrition Institute, South Asia Programme and one UGC major research project on Site specific nutrient management for enhancing rice productivity and restoring soil health in Veeranam command area, Tamil Naduas Principal Investigator. He has guided 5 M.Sc. (Ag) students. He has published 11 research articles in International journals, 19 research articles in national journals. He is a member in several scientific and professional bodies.

Dr. Muthiyan Pandiyan, Dean, Agricultural College and Research Institute, TNAU, Eachankottai, Thanjavur and Vazhavachanur, Tamil Nadu, India. He has completed his Ph.D in Plant Breeding and Genetics from Tamil Nadu Agricultural University, Coimbatore, Tamil Nadu, India. He is a notable *Vigna* breeder in national and international level. He is being developed MYMV durable resistant donor in mungbean with ricebean cross combination. He developed and released many varieties in pulses now all are ruling blackgram varieties VBN5 to VBN 8 and VBN3 and VBN4 greengram and VBN3 redgram popular among farmers, VBN7 in blackgram is national donor. Blackgram VBN 6 and VBN 8 occupied more area under cultivation. He has written ten books and 120 research articles both national and international journals. Apart from varieties, many technologies have been developed and released specifically says wild boar repellent for farming communities. He has guided more than five postgraduate and five Doctorial students and he published research and review articles at national and international reputed journals like PLOS one, Crop evolution and Plant Genetic Resources, Applied genetics and Molecular Biology.

Dr. M. Yuvaraj, Ph.D., currently working as Teaching Assistant, Agricultural College and Research Institute, Vazhavachanure, Tiruvannamalai. He completed his undergraduate degree in Adhiparasakthi Agricultural College, Kalavai during the year 2005-2009 and he was awarded International Panacea Limited (IPL) Fellowship, New Delhi and also received Dr.T.Balakrishnan Gold medal for his Post-graduate degree in Soil Science and Agricultural Chemistry at Agriculture College and Research Institute, Madurai during the year 2009-2011. He was awarded UGC-Rajiv Gandhi National Fellowship (RGNF) for his research in Nanotechnology in the field of Smart Delivery System of Zinc in Rice for doctoral research in the Department of Soil Science and Agricultural Chemistry at Tamil Nadu Agricultural University, Coimbatore during the year 2011- 2014. He worked nearly 4 years as an Assistant professor in Adhiparasakthi Agricultural College, Kalavai, Vellore at the time he got Best Teacher award during the year 2018. He published 5 international and 12 national research articles, 10 book chapters, 3 undergraduate books 4 review articles, and 16 popular articles. Besides, he serves as a reviewer and editor in various national, International journals.

Dr. C. Sivakuamr is working as Associate Professor in Agronomy at Agricultural College & Research Institute, Vazhavachanur, Tamil Nadu Agricultural University. He endowed with 15 years of experience in Agricultural Research, Education and Transfer of Technology. Through significant contribution in the field of Agronomy, he was awarded Tamil Nadu Government Chief Minister's Award during 2008 for best Tamil book, MASU award for Best Agricultural book in Tamil on "Technologies in Agricultural and Allied activities for Self Entrepreneurship Development" during 2005, Best paper and Research Award at Biennial Conference on Eco-friendly Weed Management options for sustainable Agriculture by Indian Society of Weed Science, UAS, Hebbal, Bangalore during 2001, Associated in obtaining best KVK award for Sirugamani during 2006, and Best worker award at 67th Republic day celebration during 2016 for involvement and efforts in transferring agricultural technologies to the farmers of Krishnagiri District from the District Collector, Krishnagiri.

He has wide experience in National Research Scenario and presented research papers / participated in Conferences and presented 58 research papers at national level. He has served in various scientific positions and handled project in State Balanced Growth Fund (SBGF), Remote sensing scheme and Venture Capital Scheme to the tune of 50.66 lakhs. He has developed the chemical and non - chemical technologies in weed management in direct seeded rice and it is useful for the farmers for economic weed management in rice. He also developed the different establishment technologies and foliar spraying of different phosphrous in redgram for benefit of farming community and these demonstrated to farmers,

Department officials and UG students during their exposure visit and monthly zonal workshop with demo plots.

More than 250 FLD and OFT's in different crop production technologies and participated actively in related trainings, field day, exhibitions and farm field school under ISOPOM, ICDP, NADP, NHM, ATMA schemes. Conducted / organized trainings on SRI technology in paddy, pit method of sugarcane cultivation, forage crop production, production of rice fallow pulses, micro irrigation, organic farming and precision farming in vegetable crops. He has around 144 publications to his credit including popular articles, full papers in national and international conferences and 21 Research Papers published in peer reviewed National and International Research Journals. He wrote two books in ISBN rated. During his career, he has guided six under graduate students in the field of weed management and foliar spraying in black gram and green gram respectively. He is a Life Member in several scientific societies.

Dr. A. Krishnaveni is an Environmental Scientist with 10 years of experience in Research, Education and Extension and 9 years of experience in different ICAR – NATP, UGC, and International Green Tech Projects as a senior research scholar. She has secured ICAR-Junior Research fellowship award for M.Sc (Agri.) in Plant Pathology by ICAR, New Delhi and UGC – Senior Research fellowship award for Ph.D (Envir. Sci.) by University Grants Commission (UGC), New Delhi. Best Research paper in National and International during 2003 and 2014. Best worker award from the District Collector, Krishnagiri at 67th Republic day celebration during 2016 for empowering the farm women through different organic inputs production technologies.She has handled research project as a principle investigator in Venture Capital Schemes on production of different organic inputs, vermicompost, panchagavya, VAM, Vermiwash and multiplication of earthworms and then given training to famers from different regions. She has handled organic farming technologies in different crops. Now, she is handling different environmental science courses to B.Sc (Agri.) students.

She has handled different research projects and published 27 full papers in national and international seminars and conferences with 14 Research Papers published in peer reviewed National and International Research Journals in addition to important agricultural technologies to AIR and DD news. She has guided six under graduate students in the research area of kitchen waste management, sugarcane trash composting and vemicomposting of different crop residues.

Soil Resources Inventory Management

V. Arunkumar

Department of Soil Science & Agricultural Chemistry
Agricultural College & Research Institute, Vazhavachanur
Tamil Nadu Agriculural University
Coimbatore, Tamil Nadu

M. Pandiyan

Agricultural College and Research Institute
Eachankottai, Thanjavur and Vazhavachanur
Tamil Nadu Agriculural University
Coimbatore, Tamil Nadu, India

M. Yuvaraj

Agricultural College and Research Institute
Vazhavachanure, Tiruvannamalai
Tamil Nadu Agriculural University
Coimbatore, Tamil Nadu, India

C. Sivakumar

Agricultural College & Research Institute, Vazhavachanur
Tamil Nadu Agriculural University
Coimbatore, Tamil Nadu, India

A. Krishnaveni

Environmental Scientist
Tamil Nadu Agriculural University
Coimbatore, Tamil Nadu, India

NEW INDIA PUBLISHING AGENCY

New Delhi – 110 034

NEW INDIA PUBLISHING AGENCY

101, Vikas Surya Plaza, CU Block, LSC Market
Pitam Pura, New Delhi - 110 034, India
Email: info@nipabooks.com
Web: www.nipabooks.com

For customer assistance, please contact
Phone: + 91-11-27 34 17 17 Fax: + 91-11- 27 34 16 16
E-mail: feedbacks@nipabooks.com

ISBN : 978-93-90175-92-5

Composed and Designed by NIPA.

Preface

Earth needs to be nurtured with mother's care because earth gives everything for sustaining life. Soils that form one of the most precious natural resources of earth. For sustainable agricultural production it is necessary to focus attention on the soil and climate resource base, current status of soil degradation and soil based agro technology for optimizing land use. India lives in villages and agriculture forms the back bone of the country's economy.

This book aims to provide a thorough understanding of the concepts in soil and resource inventory - Concepts - Standard soil survey - Scope and objectives - Soil systematic - Soil mapping units - Methods and types of soil survey - Soil maps. Soil Classification - Modern Soil Taxonomy - USDA System - Diagnostic horizons- Soil orders - Soils of India and Tamil Nadu. Soil survey reports-Soil Survey Interpretations - Land Capability Classification - Soil and Land Irrigability Classification Storie's Index Rating - Fertility Capability Classification- Land suitability for field crops, horticultural crops and forest trees - Land Use Planning concepts and objectives.

The book has been compiled in order to provide sound knowledge and to improve the technical competence of the Bachelor Degree of Agriculture students based on Undergraduate curriculum. The topics and contents have been presented to make it easy understanding for the students.

Authors

Contents

1

Introduction

1.1. Soil Resource Evaluation

Soil is the essence of life on the planet Earth. It has sustained humanity and human civilizations through five functions (Karlen et al. 1997) *viz.,* i) sustaining biological activity, diversity and productivity; ii) regulating and partitioning water and solute flow; iii) filtering, buffering, degrading, immobilizing and detoxifying organic and inorganic materials, including industrial and municipal byproducts and atmospheric decomposition; iv) storing and cycling nutrients and other elements within the earth's biosphere; and v) providing support to socioeconomic structures and protection for archaeological treasures associated with human habitation. Reeling under the pressure of increasing food, fodder, feed, fibre and fuel production, soil has been used as a medium of plant growth with considerable reliance on external supply of major nutrients, irrigation water, plant protection chemicals, *etc*. Although soil, a product of millions of years of weathering, has supported various forms of terrestrial life, its sustainable management holds the key for meeting the basic requirements of burgeoning population and development commitments. So intense has been the pressure of increasing food, fodder, feed, fibre and fuel production that the soil has been exploited often exceeding its carrying capacity. The soils are increasingly used for agricultural and non-agricultural uses, such as disposal of farm, urban and industrial wastes, ensuring environmental safety *etc*. This is possible by the judicious use of this resource.

A clear and intimate knowledge of the kinds of soils and their extent of distribution are essential prerequisites in developing rational land use plans for agriculture, forestry, irrigation, drainage, *etc.* soil resource inventory or mapping provides an insight about their potentials and limitations. It is therefore imperative to prepare an inventory of this resource for developing optimum land use plan and conservation plans.

1.2. Soil Resource Inventory

The process of determining the pattern of the soil cover along with potentialities and limitations, characterizing it and presenting it in understandable and interpretable form to various consumers. Soil resource inventory should be utilitarian, demand driven and cost effective way of improving the overall wealth of society.

2

Soil Survey

Soil survey is essentially a study and mapping of soils in their natural environment. It is the systematic examination, description, classification and mapping of soils of an area.

2.1. History of Soil Survey in India

The original investigations on Indian soils were carried out by Voelcker dates back to 1893 and by Leather to 1898.They classified the soils of the country into four major groups, namely the Indo-Gangetic alluvium, the black cotton soil or regur, red soil and laterite soil. Schokalskaya (1932) has published a soil map of India based on the Russian concept which describes 16 soil groups such as climate,vegetation, soil-forming materials, salinity, alkalinity, swamps and peats (Bhattacharyya et al. 2013). In 1935,Wadia and his co-workers collected a soil map of India with significance on geological formations and classified the soils as red, black (regur), laterite and lateritic soils of Peninsular India and also Indo-Gangetic Plains (IGP) which includes delta, desert, bhabar, terai and alkali soils. Vishwanath and Ukil (1943) published a soil map of India by placing the soils in different climatic zones for the first time. This was prepared by Imperial Agricultural Research Institute, New Delhi, and published in the year 1943. There were 17 kinds of soils identified, *viz.,* deep black soils, black clay soils, black loamy soils, black sandy soils, red sandy soils, brown sandy soils, red and yellow soils, dark reddish brown soils, mixed soils (white and yellow), alluvium (Indus, Gangetic, Brahmaputra), coarse alluvium, soils of swamp lands, calcareous soils, usar soils, kallar soils, coastal alluvium soils and soils of Great Himalayan Ranges.

Soil survey works were started in 1954, in most of the states as a part of their agricultural development programme, and Bihar (Soil Survey and Land use Planning Scheme, Sabour) was pioneer. In 1954, the soil survey work was started under the leadership of Dr. P. P. Jha following the conventional procedures in Bihar (including Jharkhand) to delineate the soil associations based on geology, relief, physical and chemical characteristics of soils for the first time in India by the then Soil Survey Scheme at Sabour (Mishra et al. 2001; Mishra 2015, 2016). Meanwhile, the National Atlas Organization, Kolkata, prepared a soil map of India in 1957 classifying Indian soils into six major groups and 11 broad types. Murthy and Pandey (1983) applied the working principles of the USDA 7th Approximation in classifying the Indian soils with available data on soil classes and thus used the US system for the first time at great soil group level, which was transformed into a soil map of India on 1:6.3 million scale. Mishra et al. (1994) attempted to propose "Fluvisols" as the 13th Order in USDA Soil Taxonomy for active floodplain soils frequently occurring in different parts of India and elsewhere. Mishra (2015, 2016) proposed a framework of Indian system of soil classification mainly based on land use options.

2.1.1. Establishment of Soil Survey Organization in India

Subsequent to the recognition of soil survey as a National Priority in 1947, a need was felt for creating a centralized information warehouse to assimilate, verify and disseminate information on the nature, extent and distribution of soils in the country. Consequently, the government launched All-India Soil Survey Scheme in 1956, which expanded in 1959 as the All India Soil and Land Use Survey Organization (AIS&LUS). Integrating the effects of climate, vegetation and topography, 16 major and 108 minor soil regions were identified and brought under 27 units by Raychaudhari et al. (1963). Later, a revised soil map of India was generated with 23 major soil groups under FAO/UNESCO's scheme on World Soil Map project, and subsequently this map was refined with 25 broad soil classes represented on a 1:7 million scale map (Bhattacharyya et al. 2013). In 1969, the AIS&LUS was divided into two wings, one being under the Indian Council of Agricultural Research (ICAR). It was later reconstituted as a Directorate through a Presidential Notification in 1973. The Directorate was accorded the status of a Bureau in 1976 and was renamed as National Bureau of Soil Survey and Land Use Planning (NBSS&LUP) with its head quarters at Nagpur, Maharashtra. Subsequently, the soil resources have been identified and classified through soil survey by various agencies such as the department of agriculture, soil conservation, forestry and irrigation, the All India Land Use and Soil Survey Organization, Government of India, Department of Agriculture and the National Bureau of Soil Survey and Land Use Planning of the ICAR. Dr. S. P. Raychaudhuri, Chief Soil Survey Officer carried out pioneering.work in the collection and collation of published and unpublished information on soils and their classification. It was later published in the "Final Report of the All-India Soil Survey Scheme". More than 2000 soil series were identified (Raychaudhari et al. 1963). The National Bureau also brought out a book on "Benchmark Soils of India" to coincide with the 12th Congress of the International Society of Soil

Science in 1982. Since then, National bureau (NBSS&LUP) used to bring out state, district and taluk level maps with the scale of 1:250,000, 1:50,000 and 1:10,000/1:5000, respectively, for efficient land use planning.

1846 - Geological Survey started

1880 - Dept. Of Agriculture formed

1889 - Dr . J.A. Volecker visited India

1893 - Central and provincial Agrl. Dept. started soil studies for growing of crops

1904 - Dr. Leather visited India. Grouping of soils as Alluvial, Black cotton, Red and Laterite

1928 - Royal commission on Agriculture was started. Dokuchaiev's concept of Classification introduced. Pre Irrigation Survey was conducted at lower Bhavani and Cauvery Soil fertility Survey at Cauvery, Godavari and Krishna

1956 - Ministry of Agriculture, Government of India set up the All India Soil Survey and Land Use survey organization for taking up standard soil survey In the country. This was expanded in 1959. It had four regional centres based on the soil type *viz.,*

Alluvial soil	-	Delhi
Black soil	-	Nagpur
Red and laterite soil	-	Calcutta
Red and laterite soil	-	Bangalore

These regional centres collaborate with the state centres

1976 - A new Directorate of National Bureau of Soil Survey and Land Use Planning (NBSS & LUP) was organized for imparting training in soil survey. This is functioning under ICAR. The head quarters are at Nagpur with six regional centres at Nagpur, Delhi, Calcutta, Bangalore, Udaipur and Jorhat. After the formation of NBSS & LUP, the All India Soil Survey and Land Use Organization were renamed as All India Soil and Land Use Survey.

2.1.2. History of Soil Survey in Tamil Nadu

- Soils of Cauvery Delta was studied for their fertility status (1912)
- Soils of the irrigation projects including Lower Bhavani project, Toludur Project, Cauvery – Mettur projects (1934-36) were studied
- Soil mapping of individual taluks of Tamil Nadu (1965-1986)

- Studies on Coastal soils of Tamil Nadu (1980-84)
- Soil resource mapping of Tamil Nadu jointly by NBSS&LUP and State Soil Survey Organization (1994-97).

2.2. Importance of Soil Survey

Soil is the most valuable life supporting natural resource and must be properly used. Improper use results in degradation consequently food security questioned. Rational planning in land use ensures effective exploitation of soils on its potentialities and limitations. It is necessary to prepare an inventory of this resource so that, optimum land use and conservation plans can be developed. Such plans will be useful while preparing National policies and its execution. It is also useful at other levels *viz.,* state, town, farm, watershed development, Agricultural Research Station, Research laboratories etc.

2.2.1. Uses of Soil Survey

Soil surveys are of great significance to any nation as they provide inventory of soil resources. Major uses of soil surveys are

i) They provide information for the development of land use plans for both arable and non-arable lands and for predicting.

ii) They help in predicting the adaptability of identified soils to various uses and also their behavior and productivity under defined sets of management practices.

iii) Soil resource inventory helps in recognizing the areas having constraints like salinity, alkalinity, acidity, erosion, waterlogging, flooding, etc. and in taking suitable measures for their management.

iv) Soil information generated through soil surveys is useful for land settlement, tax appraisal, locating and designing highways, airports and other engineering structures.

Soil surveys provide information about the soils of a country and form the basis for land use planning.

2.3. Standard Soil Survey: Its Scope and Objectives

Soil survey comprises a group of interlinked operations involving

- Interpretation of satellite data for delineating land for units.
- Field work to study the important characteristics of soils and the associated external land features such as landform, natural vegetation, slope etc.
- Laboratory analysis to support and supplement the field observations

- Correlation & classification of soils into defined taxonomic units.
- Mapping of soils

The mapping of soils is thus the identification, description of different kinds of soils, based on direct field observations, confirmed by laboratory data and supported by such conditions, as physiography or landforms, climate and vegetation, and delineation on a standard topographical base map. The different soils observed within a survey area are grouped into a limited number of units on the basis of their comparable physiography, morphology and physico-chemical properties. Such units of soil may occur on a landscape in a repeated manner and are delineated on the base map as map units. These units are given specific symbol, colour and name. The choice of symbols depends on the need and purpose of soil survey. A soil map may contain several map units which are the reflection of different soils delineated and given in the form of a key. An organised listing of map units is called a legend. A legend, given description of each map unit, is basic for understanding of mapped soils and forms a link between the soil map and accompanying report

- Soil survey interpretations, that is predictions about the potential of soils for alternate uses and management Transfer of technology from research stations to farmers fields through soil taxa.

Objectives of soil survey

- Prompt application of new discoveries in soil and crop management
- Determining the potential distribution and adaptability of individual crops and soil management practices
- Planning agricultural research and extending its result
- Developing rural land classification and public land management
- Planning engineering works, highways and airport and those for flood control, drainage, irrigation and conservation.
- Correlating soil conditions in the country with those of other countries and trying the discoveries on our soils.

Purposes of soil survey

Fundamental

Soil survey helps in expanding our knowledge and understanding of different soils, with regard to their properties, genesis and classification for sustainable development.

Applied

- Soil survey and soil maps help in making predictions about the behavior of different soils for agriculture, forestry, engineering, urban development etc.
- Transferring technology by correlating the characteristics of soils of known behaviour and predicting their adaptability to various uses and productivity under defined set of management practices.
- Providing information needed for developing optimum land use plans and for bringing new areas under irrigation and drainage net works
- Delineating the degraded soils, such as saline - alkali, water logged or flood prone, eroded *etc.* and suggesting soil and water conservation measures.
- Delineating disease infested areas and may provide indirect help in controlling the diseases.

2.4. Soil Systematics: Pedon, Poly Pedon, Control Section

Soil profile

The vertical cross section of soil exposing different horizons of a soil individual.

It is a two dimensional feature. Profile exploration is done upto the parent material or 2 m depth whichever is earlier. The vertical cut is made upto water table in the case of waterlogged soils. Width of profile ranges from 1 m to several metres.

The description of entire pedon or a sample within it, should record the kinds of layers, their depth and thickness and the properties of each layer. Horizons or layers are studied in both horizontal and vertical dimensions.

For selecting a profile site, the following points should be considered.

- It should be typical of the soil taxa (soil series or soil family) within the mapping unit
- It should be away from a tree, an irrigation channel / ditch / river / human settlements, road etc., as these prevent the normal development of a soil
- If possible, a virgin area should be preferred.

Soil profile – Nomenclature of soil horizons

- Master horizons - O, A, E, B, C,R
- Subordinate horizons - t, a, i, e, o, b, e, …..etc.
- Transitional horizons - AE, EA, EB, BE ….etc.

Soil Profile Description

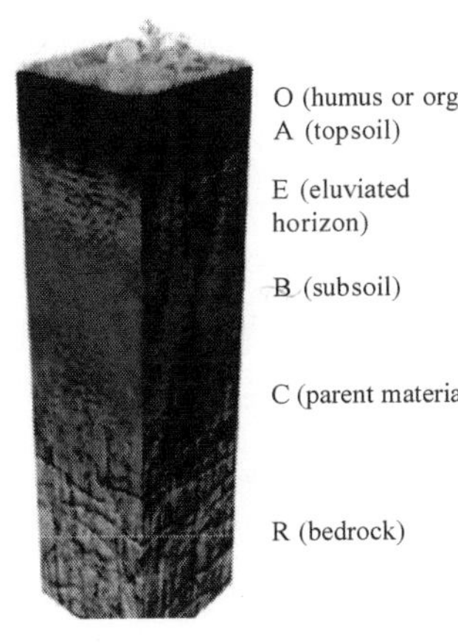

O- Horizon Surface litter, partially decomposed organic matter

A- Horizon Mineral horizon and top soil

E- Horizon Zone of leaching, material move downward

B- Horizon Iron, aluminium, humic compounds are accumulated

C- Horizon Weathered parent material

R- Horizon Bedrock

Designation of Master Horizons

The capital letters O, L, A, E, B, C, R, M, and W represent the master horizons and layers of soils. These letters are the base symbols to which other characters are added to complete the designations.

Horizon	Criteria
O	Layers dominated by organic material.
L	Limnic horizons or layers include both organic and mineral limnic materials that were either (1) deposited in water by precipitation or through the actions ofaquatic organisms, such as algae and diatoms, or (2) derived from underwater and floating aquatic plants and subsequentlymodified by aquatic animals.
A	Mineral horizons that have formed at the surface or below an O horizon. They exhibit obliteration of all or much of the original rock structure1 and show one or both of the following: (1) an accumulation of humified organic matter closely mixed with the mineral fraction and not dominated by properties characteristic of E or B horizons (defined below) or(2) properties resulting from cultivation, pasturing, or similarkinds of disturbance.
E	Mineral horizons in which the main feature is the loss of silicate clay, iron, aluminum, or some combination of these, leaving a concentration of sand and silt particles. Thesehorizons exhibit obliteration of all or much of the original rock structure
B	Horizons that have formed below an A, E, or Ohorizon. They are dominated by the obliteration of all or muchof the original rock structure and show one or more of thefollowing:1. Illuvial concentration of silicate clay, iron, aluminum,humus, carbonates, gypsum, or silica, alone or incombination;2. Evidence of the removal or addition of carbonates;3. Residual concentration of oxides;4. Coatings of sesquioxides that make the horizonconspicuously lower in color value, higher in chroma, orredder in hue, without apparent illuviation of iron;5. Alteration that forms silicate clay or liberates oxides, orboth, and that forms a granular, blocky, or prismatic structureif volume changes accompany changes in moisture content;6. Brittleness; or7. Strong gleying.

C	Horizons or layers, excludingstrongly cemented and harder bedrock, that are little affected by pedogenic processes and lack the properties of O, A, E, or B horizons. Most are mineral layers. The material of C layers may be either like or unlike the material from which the solum has presumably formed. The C horizon may have been modified, even if there is no evidence of pedogenesis.
R	Strongly cemented to indurated bedrock.
M	Root-limiting subsoil layers consisting of nearly continuous, horizontally oriented, human-manufactured materials
W	A layer of liquid water (W) or permanently frozen water (Wf) within the soil (excludes water/ice above soil)
O	Organic horizon of mineral soils
O_a	Partly decomposed organic matter
O_e	Highly decomposed organic matter
A	Mineral horizon (A) formed at or near the surface with well mixed organic matter
A_h	Uncultivated with high organic matter (> 1%)
A_p	Mixed by ploughing or other disturbances
A_g	Partially gleyed due to intermittent waterlogging, rusty mottles along root channels
E	Eluvial horizon of low organic matter content from which clay and humus have moved to lower horizons
E_g	Like A_g above but overlies a B_t horizon
B	Subsurface horizon showing typical colour, texture, structure etc., due to illuviation of material from the overlying horizons and the weathering of parent material.
B_y	Accumulation of gypsum
B_z	Accumulation of salts
B_g	Partially gleyed, blocky or prismatic structure; variable black MnO_2 mottles
B_h	Translocated organic matter with some Fe and Al
B_s	Enriched with sesquioxides usually by illuviation; orange to red in colour
B_t	Accumulation of translocated clay as evidenced by clay coatings on ped faces
B_w	Alterations of parent materials by leaching, weathering and structural reorganization
C	Parent material, excluding bed rock from which solum is believed to have formed
C_k	Accumulation of secondary $CaCO_3$ as concretions or coatings by 1 % or more (also occurs as $A_{k,}$ B_k)
C_y	Secondary accumulations of $CaSO_4$ or gypsum crystals
C_m	Continuously cemented, other than by a thin iron-pan
C_r	Weak consolidation but dense enough to prevent root penetration
C_x	Fragipan characteristics; dense but un-cemented; firm when dry and brittle when moist
D	Bed rock without decomposed by soil forming processes

Subordinate Distinctions

Lowercase letter symbols to designate subordinate distinctions within Master Horizons. Ex: b-buried soil horizon; d-dense unconsolidated materials.

Designate Subordinate Distinctions Horizons

Letter	Distinction
a	Highly decomposed organic material
b	Buried genetic horizon
c	Concretions or nodules
co	Coprogenous earth
d	Physical root restriction
di	Diatomaceous earth
e	Organic material of intermediate decomposition
f	Frozen soil or water
ff	Dry permafrost
g	Strong gleying
h	Illuvial accumulation of organic matter
i	Slightly decomposed organic material
j	Accumulation of jarosite
jj	Evidence of cryoturbation
k	Accumulation of carbonates
m	Cementation or induration
ma	Marl
n	Accumulation of sodium
o	Residual accumulation of sesquioxides
p	Tillage or other disturbance
q	Accumulation of silica
r	Weathered or soft bedrock
s	Illuvial accumulation of sesquioxides and organic matter
ss	Presence of slickensides
t	Accumulation of silicate clay
u	Presence of human-manufactured materials (artifacts)
v	Plinthite
w	Development of color or structure
x	Fragipan character
y	Accumulation of gypsum
z	Accumulation of salts more soluble than gypsum

Lithological Discontinuity

When two or more genetically unrelated (contrasting) materials are present in a profile as in the case of alluvial or colluvial soils, then the phenomenon is known as lithological discontinuity. This is indicated by the use of Roman letters as prefixes to the master horizons.

Soil horizon

A layer of soil or soil material approximately parallel to land surface and differing from adjacent layers in physical, chemical and morphological properties.

Soil Monolith - A vertical section of a soil profile removed from the field and mounted for display or study.

Soil Solum - Upper and most weathered part of the soil profile representing A and B horizons

Soil Sequum - Continuity and discontinuity along different axis of soil profile

Soil section – 9 inches thickness of soil profile.

Regolith – Inclusive term for all the loose materials above bed rock

Ped – Unit of soil structure

Intergrade horizon – No clear cut boundary between the horizons

Standard horizon – Clear cut boundary between the horizons

Horizonation – Pathways leading to the development of horizons

Haploidisation- No horizon formation due to pro isotropic processes (Weathering of sedimentary rocks).

Pedon

A pedon is the smallest volume that can be called "a soil ". At the same time, it must be large enough volume of soil to be observable and to exhibit a full set of horizons. It is three dimensional; roughly hexagonal in shape. Its area ranges from 1 to 10 m^2 upon soil variability. It is a unit for soil sampling.

Polypedon

These are the group of contiguous (adjacent and in close contact with) similar pedons bounded on all sides by not soil or by pedons of unlike characters. It is a real physical soil body which has a minimum area of more than one sq. km and an unspecified maximum area. The polypedon are the real soil bodies that we classify into series and higher categories.

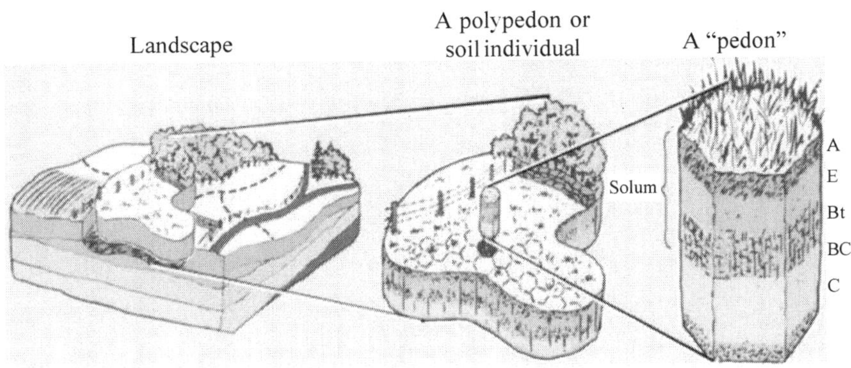

Soil individual

A soil individual is a soil body that may be defined in terms of profile features whose arrangements and combinations over a geographic area are unique.

Control section

It is a portion of the soil profile delimited in terms of an arbitrary depth or depth range in centimeters or inches or the part of the soil on which classification is based. The thickness varies among different kinds of soils, but for many it is 40 or 80 inches (1 or 2 meters). The limits of soil moisture control section (upper and lower) are determined by the soil depths to which a dry soil at wilting point (kPAe" 1500, but not air dry) is moistened by:

- 2.5 cm after 24 hour
- 7.5 cm after 48 hour

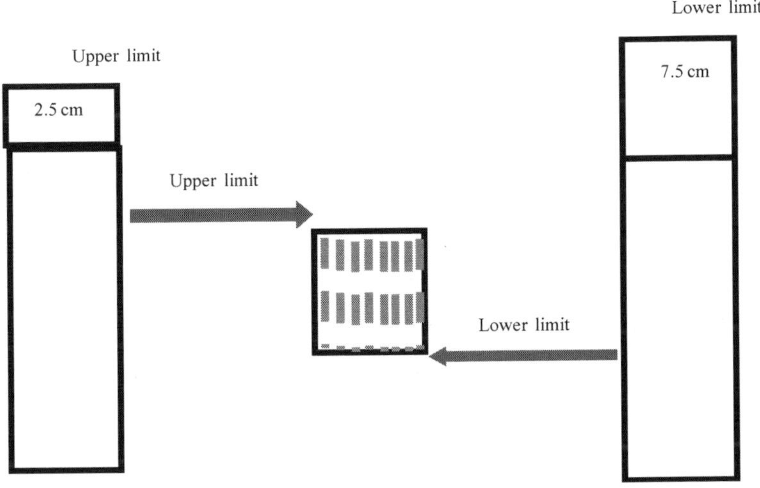

The upper and lower boundaries of the soil moisture control section extend to different depths, depending on the texture as under:

- From 10 to 30 cm in fine (clayey;fine loamy or fine silty) soils
- From 20 to 60 cm in medium (loamy or coarse loamy) soils and
- From 30 to 90 cm in coarse (sandy) soils

Soil three dimensional sequence

Soil is recognized as three dimensional natural bodies having length breath (area) and depth.

Vertical: Soil surface to depth where not affected by pedogenic processes (2 axis)

Lateral: Succession of contiguous soil bodies in horizontal direction (X and Y axis)

2.5. Soil Mapping Units Soil Series, Soil Association, Soil Complex Variants

Soil mapping units

The classification of areas having similar soil components. It differs from other mapping unit in some of the soil characteristics depending upon the purpose. The following mapping units are being used in different kinds of soil survey.

Soil series

It is a mapping unit consisting of a group of soils having similar pedon profile characteristics. Basically the morphological characters are taken into account for characterizing a soil series. The characters considered for characterization of a soil series are

a) Arrangement of soil horizon
b) Thickness of the individual horizon
c) Solum depth
d) Soil colour
e) Soil structure
f) Consistency
g) Presence of clay films
h) Soil pores
i) Root distribution
j) Development of surface cracks

By using profile characteristics we differentiate the soils in the field. The physical and chemical characteristics are not taken for mapping. The environmental characters like slope, topography, stoniness, rockiness, soil erosion and the surface layers are not taken in to account for characterizing the soil.

Each soil series is named after the name of a village, town or famous temples in which it is established. For establishing a soil series a minimum area of 2000 acres in required. The soil series is represented on the soil map by three letter symbol bearing the name of the soil series.

eg:

Plm	- Pilamedu	Tyr	- Turaiyur,
Igr	- Irugur	Klt	- Kalathur
Mdk	- Madukkur	Tlk	- Thulakkanur
Vgg	- Vyalogam	Pvd	- Palaviduthi
Klk	- Kallakudi	Plr	- Puvalur
PKt	- Pattukottai	Omd	- Omandur

Soil Association

It is a mapping unit consisting of two or three soil series occur in a repeated geographical pattern. The boundaries of soil series can not be demarcated separately due to map scale limitation. Under these circumstances mapping unit soil association proposed.

eg: Palathurai - Pichanur soil series occur in association. This major soil series should be written first followed by the minor. Here also three letter symbol is used.

Pth - Pch (Palathurai - Pichanur)

Plm - Pth - Pch (Pilamedu - Palathurai - Pichanur)

Soil Complex

It is a mapping unit consisting of more than three series occurring together in a repeated geographical pattern. Boundaries can not be demarcated separately. It is a mapping unit in hilly terrains. It is indicated by Roman letters. The mapping units namely soil series, association and soil complex are used in the Reconnaissance soil survey.

Soil type

It refers to the surface texture of soil series. It is denoted by the name of the soil series followed by the texture of the surface layer.

eg: Plm - sandy loam

Soil phase

This is a mapping unit classified based on the environmental characters of the soil series namely slope, erosion, stoniness etc.

Type and phase mapping units are used in detailed soil survey.

Miscellaneous land types

The areas which have little or no soil to support any vegetation are named as miscellaneous land types. These areas could be used only after proper reclamation. Mostly they are non soil however, they are indicated along with soil series, associations and complexes. Following are included in the above heading.

i) Water bodies
ii) Salt pans
iii) Sand dunes
iv) Rubble lands
v) Rockout crops
vi) Ravines .... etc.

Inclusion

These are the soil not denoted in soil map. They are so small to be delineated. Normally soils which occupy less than 20 per cent of the mapping area belongs to this type.

Variants

These are soils of a established series differing in some properties of the series. They are indicated as variant of the series in reconnaissance soil series. These variants will be proposed as a new series during detailed soil survey.

Mapping Symbol

Maps use symbols to represent features on the ground. These features include roads, tracks, rivers, lakes, vegetation, fences, buildings, powerlines, administrative boundaries and the like. Given the size of a map, it is not possible to show all features that occur on the ground. Large scale maps show more details and a larger number of features. Colour plays an important part in symbols for example, blue for water features and green for vegetation.

Roads, metalled: according to importance: distance stone
,, unmetalled; do. do. ; bridge
Cart-track. Pack-track and pass. Foot-path with bridge
Bridges: with piers: without. Causeway. Ford or Ferry
Streams: with track in bed; undefined. Canal
Dams: masonry or rock-filled; earthwork. Weir
River banks: shelving; steep, 3 to 6 metres; over 6 metres
,, dry with water channel; with island & rocks. Tidal river
Submerged rocks. Shoal. Swamp. Reeds
Wells: lined; unlined. Tube-well. Spring. Tanks: perennial; dry
Embankments: road or rail; tank. Broken ground
Railways, broad gauge: double; single with station; under constrn.
,, other gauges: do. ; do. with distance stone; do.
Mineral line or tramway. Telegraph line. Cutting with tunnel
Contours with sub-features. Rocky slopes. Cliffs
Sand features: (1) flat (2) sand-hills and dunes (surveyed) (3) shifting dunes

Map scale

A map represents a given area on the ground. A map scale refers to the relationship (or ratio) between distance on a map and the corresponding distance on the ground. Map scales can be shown using a scale bar.

Based on the scale of mapping, the map is called as

Large scale maps	-	1: 5000 and less
Medium scale maps	-	1:25,000 to 1:40,000
Small scale maps	-	1:50,000 to 1: 2, 50,000
Very small maps	-	1: million and above

	Scale	Ground Distance of 1 cm on the map
Larger	1 : 10000	100 m
	1 : 25000	250 m
	1 : 50000	500 m
	1 : 100000	1 km
Smaller	1 : 250000	2.5 km
	1 : 1000000	10 km
	1 : 5000000	50 km
	1 : 10000000	100 km

Larger the scale of a map, smaller the area that is covered and more detailed is the graphic representation of the ground. Small scale maps (such as 1:250 000) are good for long distance vehicle navigation, while large scale maps (1:50 000) are ideal for travel on foot. Scales are usually shown in increments of one, five or 10 kilometres. Soil maps generally contain more than one map unit. These units are arranged in many ways. An organized list of map units is called a Legend. There are several kinds of legends. Map units have symbols, colours and names. The nomenclature may differ from one legend to another.

Taxonomic units Versus Mapping units

Taxonomic units define specific range of soil properties in relationship to the total range of properties measured in the soil. The map units and their individual delineations define areas on a landscape. Every mapping unit has more than one taxonomic unit (soil series or soil phase).

Kind of mapping units

Mapping units have been distinguished based on the amount of inclusion or impurity they contain. Five kind of mapping units have thus been distinguished and these are consociation, association, complex, undifferentiated and miscellaneous/unvisited.

Consociation

A consociation is a mapping unit with very little inclusion or impurity. It is assumed to contain the profile class after which it is named but in practice the purity of such class may range from 70% to 85%.

Association

An association is a mapping unit that contain two or more taxonomic class that are nearly equally represented and in which it is very easy to separate one profile class from the other.

Complex

Complex is a mapping unit where more than two taxonomic classes are equally represented and the components are intricate so that separation, even at large scale is difficult.

Undifferentiated

This is a mapping unit consisting of a number of taxonomic units that are merged so that separation into different units are impossible at any reasonable mapping scale

Soil survey methods

Soil survey methods have changed greatly in the recent past. Soil scientists started the soil survey using blank paper as a basemap, aerial photograph as basemap and now satellite data is used as base map.

Base Maps

The fundamental requirement of all mapping activities is a suitable base map. These base maps need to be complete in details of features and accurate in their location to enable surveyor to delineate soil boundaries more correctly and conveniently. Depending on the intensity of mapping, following types of base maps are used for soil surveys in India and in many other countries.

Cadastral maps

Cadastral maps on the scale of 1: 2640 (24" = 1 mile) to 1:7920 (8" = 1 mile) or 1: 15,840 (4" = 1 mile) in plain areas and 1:1200 ((52.8" = 1 mile) in hilly areas are used for detailed mapping. Cadastral maps show field boundaries and field or revenue survey number, however, they lack the topographical details (contours, elevators, etc.).

One advantage in using cadastral base is that the soil survey information or interpretations can be communicated to individual farmers by reference to the field survey number. The cadastral maps can be procured from the '*village administrative officer*' or concerned '*Tahsildar*'.

Topographical maps

Topographical maps are published on the scale of 1:25,000, 1: 50,000 and 1: 250,000. These are used as base maps for soil surveys in India. Topographical maps show not only physical features but also contain topographical details in the form of contours and elevation above mean sea level. These maps have reliable planimetric accurately facilitating measurement of distances and easy preparation of soil map. In India, topographical maps are prepared and published by the Survey of India, Dehradun and can be obtained from them or their regional offices.

Aerial photographs

Aerial photographs are the pictures taken by camera fitted in an aircraft and flying over the terrain at a predetermined height depending on the scale of aerial photography and focal length of camera. Aerial photographs give a bird's eye view of large areas. The aerial photographs ranging in scale from 1:8000 to 1:60000 are used in different types of soil surveys.

Relief can be perceived by stereoscopic study of aerial photographs. Relief features help in identifying various kinds of land forms which are elated to different kinds of soils. Many landforms e.g.terraces, flood plains, sand dunes, coastal plains, plateau, paleochannels, hills, valley, mountains, etc. can be recognized on the photographs from their shape, relateive height and slopes. Differences in tone or colour may also reflect soil differences.

Individual objects like trees, houses, roads, footpaths, field boundaries, lakes, river courses are imaged clearly depending on the scale of photographs. These 'land marks' serve as effective reference points or 'local control points' that facilitate a soil surveyor in orientation and navigation during the field work and in demarcation of soil boundaries of high local accuracy.

Remote sensing

Remote sensing is the science of obtaining information about objects or phenomenon in the environment through the use of sensing devices located at a distance without there being any contact between the object and the sensing device. The satellite data on 1: 25,000 to 1: 250,000 can be used for semi detailed and reconnaissance survey of a district or a region for planning.

2.6. Methods of Soil Survey

Soil surveys are being conducted for diagnosing the nature and extent of distribution of different soils, including normal and problem soils. There are two methods of survey.

1. Grid survey
2. Free survey

Soil and Landform attributes

Many soil and landform attributes such as soil texture, depth, profile development, drainage, soil moisture and temperature regimes can be observed and mapped.

General Traversing

The first step in soil survey begins with the general traversing of the area. Recent use of aerial or satellite imageries has become common as they have more details than cadastral or topographical maps. The surveyor studies the topography, geology, climate and vegetation – the factors of soil formation. After the preliminary studies, the surveyor goes to the field to study the soils and prepares a field mapping legend, based on different attributes that affect the plant growth or based on soil genesis. The preliminary legend provides the basis for soil mapping.

For carrying out the field work, the surveyor walks briskly across the fields with a base map and auger in his hands. The kind of auger used depends upon the terrain or area being surveyed. After few minutes, the surveyor stops to bore a hole in the soil in order to note the sub soil variations in different attributes / characteristics such as texture, depth, colour, moisture status, mottling, concretions, calcareousness etc., Thus he identifies soil units. He thus determines the depth of horizons and also the illuvial and elluvial horizons. He makes observation on land use, relief, erosion and then draws the same on the base map . He collects soil samples at intervals depending upon the variability in soils and intensity of mapping.

Observations recorded during traversing

Soil is composed of distinct layers called horizons. The intensity of colour in a layer denotes the amount of organic matter, lime , iron or reduction and oxidation status. The lower layers have streaks or spots of grey, yellow or red colours called mottles, which indicate impeded drainage and restricted aeration.

Texture which is the relative proportion of sand, silt and clay is determined by feel method and is confirmed by mechanical analysis in the laboratory

The depth of bed rock or compact layer is determined.

External features such as quantity of gravel, stones or rocks , slope, erosion, salinity are also recorded.

2.6.1. Grid Survey

The procedure is adopted for mapping small areas, such as a micro watershed or agricultural research station. The traverse lines are located on a grid pattern and the density of mapping is adjusted according to the area surveyed so that the number of observations per cm^2 of the final map is independent of the scale. Generally four or five observations per ha are recommended. Points of comparable observations are drawn. In this computer age, grid survey at geo referenced points (with latitude and longitude) have been found to be of great value in digitizing the database and generating several thematic maps of practical value. Such surveys are expensive.

2.6.2. Free Survey

In free survey, the surveyor chooses observation points on the assumption that changes in physiography, as such observed by aerial photo or satellite imagery interpretation and other surface features, such as colour, vegetation and land use are indicative of differences in soil characteristics.

The density of observations can be varied as the mapper concentrates on confirming the inferred boundaries and checking the uniformity of the soil within each boundary. On small scale, the inferred boundaries are often accepted as soil boundaries. On large scale, several new boundaries within the physiographic boundaries are recognized depending upon the scale of mapping.

Based on the differences in physiography, parent material and drainage class and profile development, the surveyor groups soils of the area into defined soil units, the limits of which are rigidly defined as per criteria used. For general purpose, each soil unit is built around a central concept covering a limited range in values of several soil parameters. Sometimes the soils are grouped at soil series level. A soil series is group of soils similar in their differentiating characteristics and arrangement of horizons.

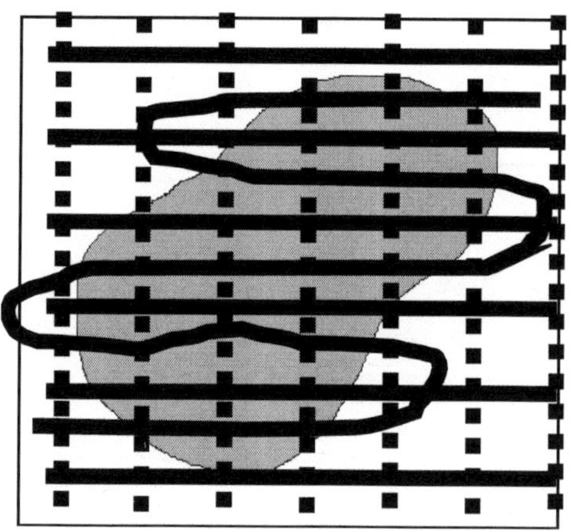

Fig. Grid survey procedure for traversing

2.7. Types of Soil Survey

1. Reconnaissance Soil Survey (RSS)
2. Detailed Soil Survey (DSS)
3. Detailed Reconnaissance Soil Survey (DRSS)
4. Semi Detailed Soil Survey (SDSS)
5. Exploratory Soil Survey (ESS)
6. Rapid Reconnaissance Soil Survey (RRSS)

2.7.1. Reconnaissance Soil Survey (RSS)

The reconnaissance survey is undertaken to prepare resource inventory of large areas. It identifies broadly, the kinds of soils and their extent of distribution. It enables to assess broad potentialities of soils and recognition of areas of promise that are suitable for intensive agriculture and that requiring priority treatment for amelioration. It is low intensity soil survey.

The survey relies on aerial photo or remotely - sensed data interpretation to develop soil physiographic relationship. The soil map provides information needed for broad or regional level planning in respect of land use. This kind of mapping proceeds detailed soil survey of potential areas.

Scale of mapping	:	1:250000 to 1:100000 / 1:50000
Base map	:	Toposheets, aerial photographs or satellite imageries
Intensity of observation	:	Soil profile is examined at 3-6 kms or even shorter, depending upon soil heterogenity. Auger sampling is done at an interval of 2.5 to 1 km.
Mapping units	:	Association of soil series, families and great groups
Purpose of RSS	:	It provides information needed for broad land use planning and agricultural development. The information obtained is used for Taluk, district and state level planning. It precedes detailed soil survey and mapping
Target	:	1000 acres / day / soil survey party (SSP)

Steps involved in Reconnaissance soil survey Pre field study using base map

The toposheets for the particular area to be surveyed to be obtained and information regarding geology, land use of area and other particulars are gathered. This gives some information before field visit.

Rapid traversing of the area (by jeep)

Soil survey party (SSP) takes up rapid travel in the entire main road to study the distribution of different kinds of soil. During the travel, soils will be examined at larger intervals. The heterogeneity of the soils and distribution are known during this stage. SSP will be in a position to know tentatively the number of soil series distributed in a study area.

Fixation of soil legends

Five to ten profiles will be opened for each group of soil identified in step *viz.,* red soil, black soil. The profile characteristics will be examined and range of characteristics for each group will be fixed.

Legend of RSS

- Location of the study area
- Physiography
- Geology
- Tentative soils series

Profile morphology of the tentative soil series

A graph comprising the climate, geology, land use and kinds of soil with range of characters will be prepared and circulated among soil survey party for use. This forms the legend which is amenable for modification if any during the course of regular field work.

Regular field work

Soil survey party divide among them and initiate the field work from different points of the study area. They delineate and indicate the boundaries of soil identified in the legend. The mapping of the boundaries is done by examining the profiles are 4 to 6 km intervals an auger sampling at an interval of 1-2 kms. If a soil survey party comes across a new kind of soil during the field work other than that was fixed in the legend, then a new name will be proposed and the boundaries will be delineated by profile examination.

The mapping units used for delineation include soil series, soil association and soil complex. Areas which are accessible will be examined and the boundaries will be extended to the inaccessible areas. This process is known as Extrapolation.

Collection of soil samples

After completion of field work, soil samples are collected from the representative profile for each soil series for analysis. Monolith soil series may be collected from representative profile which may be used for soil correlation studies. In Reconnaissance soil survey, subsoil characters are given prime importance.

2.7.2. Detailed Soil Survey (DSS)

This type of survey is undertaken in priority areas, such as pilot projects, agricultural research stations, micro water sheds and areas in urban development. The traverse lines are on grid pattern. The survey enables identification of soil units up to phases of series for planning development of individual parcels of land. The resulting soil map provides sufficient information for interpretation of various kinds of soils and for understanding their pedogenic evolution. Such a survey is very time consuming and expensive. The detailed soil survey is of two type's e.g. low and high intensity.

Scale of mapping	:	8" = 1 mile; 16" = 1 mile
Base map	:	Cadastral map / village map
Intensity of observation	:	Soil profile is examined at every one ha and auger sampling at an interval of 0.25 to 0.5 km
Mapping units	:	Soil type, soil phase
Purpose of DSS	:	It provides information needed for village, farm and block level planning
Target	:	200 acres / day / soil survey party (SSP) – 40,000 acres per year

Steps involved in Detail Soil Survey

1. **Initial traverse of the study area** – A rapid traverse will be taken up to know the distribution of soil types and soil phases of soil series by soil survey party. The list of types and phases of the soil series is prepared and used as a guide during the regular field work.
2. **Legend** – The soil legend gives a range of soil texture in the soil series, depth range and slope and erosion classes.

Regular field work – Soil survey party will start the work from a permanent point. Mapping units such as phases and types of soil survey will be demarcated during traverse. SSP cover the study area in foot while Reconnaissance soil survey in jeep. Soil Survey Party walks in a zig-zag manner. During the traverse party will examine the soil series at an interval of 0.25 to 0.5 km by auger boring and every 640 acres by profile digging. Difference between two traverse lines will be minimum of 0.25 km. Thus boundaries of the types and phases are demarcated. After completing the Detail Soil Survey, soil samples are collected from the representative soil profile from each mapping unit prescribed for Detail Soil Survey.

1. Mapping units for a Detail Soil Survey

The mapping units for a DSS are written like a formula. It consists of name of the series, surface texture, depth class and slope and erosion class.

E.g. $\frac{\text{pth-sl-}d_5}{\text{B-}e_2}$

Where,

Pth Name of the soil series

Sl	surface texture of the soil series
d_5	Depth of solum
B	Slope percentage
e_2	Erosion class

pth-sl-d_5 Soil type

B-e_2 Soil phase

Those areas which could not be delineated separately because of their smaller extent can be denoted by conventional symbols.

2.7.3. Detailed Reconnaissance Soil Survey (DRSS)

This kind of survey combines both the detailed and reconnaissance surveys as above and is undertaken for understanding distribution of basic soil classes of series families and their phases. First Reconnaissance soil survey is conducted for the whole study area and this is followed by Detailed Soil Survey in selected areas such as intensively cultivated areas, problem soil areas, wasteland development, polluted soil areas or potential areas for a particular crop variety. Ex. If the problem soils of Coimbatore are to be studied, Reconnaissance soil survey is to be undertaken for the whole area and Detailed Soil Survey in those problematic areas identified.

2.7.4. Semi Detailed Soil Survey (SDSS)

This survey comprises very detailed study of some selected strips cutting across many physiographic units for developing correlation between physiographic units and soils. Once correlation is developed and is found to be valid by random checking, the rest of the area is checked at random and soil boundaries based on physiographic units delineated. This kind of survey provides sufficient information about various kinds of soils including problematic or degraded soils. Scale of base maps (aerial photographs or satellite imagery) used is 1:50,000. Mapping unit is the association of soil series or families. The final maps are prepared on 1:50,000 scale.

2.7.5. Exploratory Soil Survey (ESS)

Exploratory surveys are not survey proper. They are usually rapid road traverse made to provide modicum of information about the area that are otherwise unknown. Scale of exploratory survey varies from 1: 2,000,000 to 1,500,000. These lead to preparation of small scale soil maps that are needed for macro level planning for varied agro -based development programmes.

2.7.6. Rapid Reconnaissance Soil Survey (RRSS)

In this survey, field mapping is done at 1:1000000 or still smaller scale using the satellite imagery. Soils are mapped by traversing representative areas. The mapping units are phases of great groups. Observations are done at an interval of 1-2.5 km.

Table. Relationship of Scale of soil Map and Frequency of field observations.

Types of Survey	Scale of base map	Area represented by 1 cm^2 on map	Distance between field observations	Frequency of observation	Mapping unit	Field Procedure and Accuracy of Soil boundaries
Rapid RSS	1: 1,000,000	10,000	10 km	1 in 10,000 ha	Phases of soil great groups	Boundaries are plotted by interpretation of remotely sensed data
RSS	1: 2,50,000	625	2.5 km	1 in 625 ha	Phases of association of soil families	Boundaries are plotted by interpretation of remotely sensed data & verified with observation at random
	1: 1,00,000	100	1.0 km	1 in 100 ha		
Semi detailed	1 : 50,000	25	500 m	1 in 25 ha	Phases of soil series / association of soil families	Boundaries in each delineation in sample areas are identified by actual traversing- others by remotely sensed data
DSS-Low Intensity	1: 10,000	1	100 m	1 per ha	Phases of soil series	Almost all boundaries are checked through traversing
DSS- High Intensity	1: 5,000	0.25	50 m	4 per ha	Phases of soil series	All boundaries are checked through traversing

2.8. Soil and Landscape characteristics Observed During Soil Survey

2.8.1 Site Features

The following site features are to be observed during the examination of soils.

i) Location and geographical landscape

Location of the profile should be indicated with reference to latitude and longitude. It should be away from roads, buildings and should represent the area. The geographical land features like basin, depression are to be indicated. Elevation also indicated.

ii) Slope

The percentage of slope and direction of the slope are to be indicated. Appropriate symbols should be used to indicate slope nature according to standard classes recognized.

iii) Climate

The various climate data like rainfall, temperature, potential evapotranspiration, and soil moisture and soil temperature are to be recorded. Here more emphasis is given for the temperature regions.

iv) Vegetation

It refers to the natural vegetation. Here land use also considered. Based on the land use it is described as forest, crop land, terraced land, pasture land, plantations etc.

v) Erosion

It refers to removal of soil material by wind and / or water. Gradations of wind erosion *viz* slight, moderate, and severe and water erosion viz., splash, rill, sheet, gully are indicated.

vi) Ground water

It refers to depth from soil surface to the body of water surface. This information is being collected by enquiring the local people and also by observing the open well in that area.

vii) Parent material

The base material from which soil is formed is to be noted. The type of rocks viz., igneous, sedimentary, metamorphic is also indicated.

viii) Drainage

The various drainage classes viz., partial, imperfect, moderate, well and excessive are noted by appropriate symbols.

ix) Presence of stones and rock out crops

The material having the dimension ranging from 75-250 mm is called stones. The classes of stoniness are very few, fairly story, stony, very stony, excessively stony and ribbed land, which are based on percentage of area covered by stores.

x) Presence of salt / Alkali

It refers to the accumulation of salts usually linked with white alkali. This is recorded by observing the white patches on the surface of the soil. This will be confirmed by laboratory analysis and grouped into classes.

xi) Moisture content of the soil

The prevailing moisture content of the profile is to be recorded

xii) Higher category level of soil

The higher category levels of soils like order, sub order and great group etc are recorded.

2.8.2. Description of Soil Profile

The soil profile must be perpendicular to sun's movement for observation and taking photograph. It should be always North-South. The length at North-South should be 2 meter and width at East-West should be 1.5 m. Depth of profile should be 2 m or up to parent rock or up to water table whichever is less. While digging profile pit the surface, sub surface soil must be kept both North and South sides in profile, the steps are provided in North – South direction, for easy and close observation. The profile pit is examined at morning on western side and at evening on eastern side. Following parameters are studied in a profile for better understanding and classification of soils.

i) Horizon symbol: In a profile first boundaries of different horizons are to be marked by observing colour difference. On marking boundaries, they (horizons) have to be named using symbols as per latest nomenclature.

ii) Horizon thickness: Each horizon thickness is to be recorded. This will indicate the development nature of the soil in that area.

iii) Colour:The colour of each horizon is recorded using Munsell colour chart both in wet and dry conditions.

iv) Mottlings: Spots or streaks of different colours (usually orange or grey) inter spread with the dominant soil matrix colour developed due to periodic reduction/ water-logged and oxidation conditions. Mottling with chroma 2 or less indicate poor drainage. Mottlings described in terms of three characteristics viz. abundance, size, contrast etc.

v) Texture: The texture of each horizon is recorded initially by feel method and later confirmed by laboratory analysis.

vi) Structure: The arrangement of soil particles in each horizon is indicated by using appropriate symbols.

vii) Consistency: It is the behavior of soil cohesion under various moisture conditions towards any force. The consistency of soil is observed and symbol of appropriate classes is indicated.

viii) Cutans / clay films: These are thin oriented clay on surfaces or in soil pores. During the field investigation it can be observed as thin shiny surfaces either on pits or in pores by using 10x or 20x hand lens. Based on thickness it is classified as thick or thin. Based on the abundance it is classified as patchy or continuous.

ix) Concretions : These are hardened materials found in soil horizons. They are formed due to secondary accumulation of soil materials around a nucleus. These are described using characteristics like abundance, size and types.

x) Soil pores: The soil pores in each horizon is indicated by characteristics like abundance and size.

xi) Presence of roots: This is described in terms of abundance and size.

xii) pH of soil horizons: The pH of each soil horizon is recorded using pH indicator papers at field. Later it is confirmed with laboratory analysis.

Examination of soil profile

A suitable site representing the normal conditions prevailing in the area is selected for profile examination. The site should be representative of a sizeable area and should be away from the field boundaries, roads or rivers. Extremes of relief are avoided in the selection of sites. Normally a pit of the dimensions of 4 ft long and 5 ft deep is dug. While digging, care sould be taken to expose the vertical face to sunlight and the soil excavated be thrown away to either side. It is important that the side to be examined is vertical and even. Any special features observed during the process if digging should be recorded. The different horizons are identified, their relative depth and features like colour, texture, structure, etc., studied in detail and recorded in the standard pattern.

Field study of soils, as they occur in different horizons forming a profile, involves characterization of the various physical and physicochemical properties that can be suitably assessed. These properties include soil colour, texture, structure, consistency, permeability and also such others like carbonates pH.

2.9. Soil Mapping

Soil mapping comprises of identification, description and delineation of different kinds of soils based on climate and vegetation of the area, confirmation through field work and laboratory data and depicting on a standard base map. The different soils observed within an area are put into limited number of groups. Each group of soils is typified by a pedon or their association. A pedon is morphologically described in the field and sampled for laboratory investigations for their characterization and classification. Such group of soils may occur on the landscape in a repeated manner and are delineated on a map as 'map unit'. The delineated units on a map will be given a specific symbol, colour and name. soil maps contain several mapping units along with a legend that give a description of each soil unit. The soil maps are prepared on different scales varying from 1:1 million to 1:4000 to meet the requirements of planning at various levels.

Geo-pedological approach to soil mapping

The Geo-pedological Approach to soil survey was developed by Zink (1988) and is essentially a systematic application of geomorphic analysis to soil mapping. This approach is a hierarchical system applied in semi-detailed studies. This approach can be used to cover large areas rapidly, especially the relation between geomorphology and soils is close. It is based on two hypotheses. The first one is, boundaries drawn by landscape analysis separate most of the variation in the soils. This will be the case if the three soil forming factors (parent material, relief, time) which can be analysed by this approach are dominant and also if the mapper has correctly interpreted the satellite data and sample areas. The interpreter must form a correct model based on the geomorphology (soil-landscape relations) and apply it correctly and consistently. If sample areas are representative, soil pattern can be reliably extrapolated to unvisited map units.

Remote sensing and Soil mapping

Before the advent of remote sensing techniques, the soil surveys were conventionally carried out at reconnaissance, detailed and detailed reconnaissance level depending upon the requirements. In all these three methods soil mapping is done after field observations. Later, aerial photographs were used as remote sensing tool in soil mapping and because of some technical limitations they could not be operationalized in soil mapping.

Use of aerial photograph in soil mapping

Among the different aerial photograph, black and white, colour infra red (IR) and colour Infra red (CIR) aerial photographs are used in soil mapping. Aerial photographs with a scale of 1: 40,000 to 1: 60, 000 for reconnaissance soil mapping and 1: 10,000 to 1: 25, 000 for detailed soil mapping are used. Aerial photographs permit 3D view through stereoscopes and hence slope, drainage pattern, natural features like hills, valleys and plains can be easily distinguished in a given geological formation. Sub divisions of landform (hlls, pediment, pediplain valley, alluvial plain etc) can be delineated using photo elements (slope, erosion, tone, texture, density of reservation, land use etc.) Physiographic units for each land form are identified.

The physiographic units are studied in detail for the soil composition. The steps involved the use of aerial photography for soil mapping is given in figure 1. Orthorectifiction has to be done if rectified aerial photograph are not used in soil mapping.

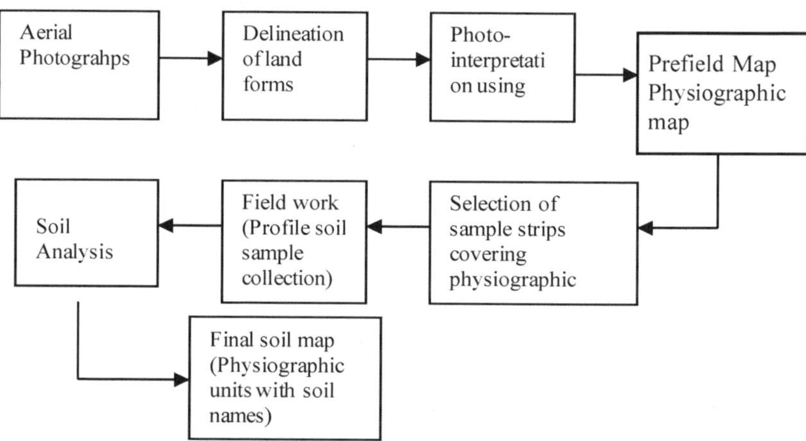

Fig. 2.1 Flow chart for soil map preparation using aerial photographs

Use of satellite data for soil mapping

Application of satellite remote sensing data from satellites for soil studies began with the launch of Landsat-1 in 1972. Remote sensing techniques have reduced field work to a considerable extent and soil boundaries are more precisely delineated than in conventional methods. While mapping the soils using remote sensing technique, the stereo data is highly useful in identification of different landforms, which have got close relationship with the soils associated with them. The stereo data from panchromatic (PAN) cameras aboard SPOT/IRS-1C/ Cartosat-1 enabled the delineation of physiographic units and soil maps derived therefrom in a better way.

Visual interpretation for soil mapping

The approach for soil mapping using visual techniques h as been summarized in Figure 2. The broad methodology for mapping soils has been explained hereunder.

Satellite data

Satellite imageries (Photographic format) and digital data are used for soil mapping. Satellite imageries are available in 1:1 million, 1: 250,000, 1: 50,000 and 1: 25,000 scales are available for generating soil maps for different levels of planning. Summer season data are preferable for soil mapping. However, under certain terrain conditions the data of monsoon in association with summer season helps to better extract information on soil types. The scale of soil map determines the sensor that has to be employed in the soil mapping.

Ancillary information

The ancillary data required for preparation of soil map consists of topographical maps, published soil maps, geological maps, reports, climatic data (rainfall, temperature etc.). Topographical map sheets are required for preparation of basemaps, ground truth collection, in the selection of sample strips.

Preliminary interpretation

The principle of 'geo-pedological approach' is employed in preparing soil map. The satellite data (False colour composite) of the study area will be interpreted for broad physiographic units on a computer screen. On the basis of available lithological information through ancillary data, the different physiographic units are attached with the geological information. The sub division of broad physiographic units will be carried out based on drainage, network, erosion, land use etc., as manifested on the satellite data. A tentative legend will be developed for the preliminary interpreted map in terms of physiography, lithology and mapping units.

Selection of soil sample areas

On the basis of preliminary interpretation, sample areas covering all mapping units in the legend are selected in the form of sample strips of suitable size. The size of sample strip is usually about 2 x 5 km covering at least 2 or more mapping units. The number of sample strips may vary depending upon the variability in lithology, physiography, vegetation cover *etc.* Through sample strips, a miniumum of 10 per cent of total study area is covered in a toposequence. This may be increased depending upon the heterogeneity in the terrain conditions. Besides, in each sample strip, random observations in the form of auger bores or minipits are taken to account for variability in the soils within the mappng unit. The selected sample strips are transferred on to the base map for field sample colletction.

Ground truth collection

Initially a rapid traverse of the entire study area is undertaken to adjust the sample strips. Subsequently, in each sample strip, soil profiles, minipits, augerbores etc., are studied along with morphological characteristics, existing land use/ terrain parameters. The information recorded systematically by noting the place of observation using Global positioning system, topographic conditions, existing land use /land cover, geology of the area, physiographic unit, natural vegetation occurring in the area and morphological characteristics of the soils. Random observation are also made in the study area to confirm the soils in different mapping units. The soil samples are collected from soil profile horizons with proper labeling to determine various soil properties later in the laboratory.

Soil sample analysis

The soil samples collected during the field work are processed after air drying of the soils. They are pounded and grinded so that soil sample passes through 2 mm sieve. These soil samples are analysed for various physical and chemical properties like pH, EC, texture, organic carbon, exchangeable cations, CEC etc., that aid in the taxonomic classification of soils and also to identify any salinity or alkalinity like problems.

Classification of soils

The soils are classified, based on the morphological, physical and chemical properties collected in the field work and laboratory analysis, following the soil Taxonomy of USDA. The scale of mapping determines the level of classification of soils. For example, the soils are classified upto association of sub-group level at 1:250,000 scale or at soil series at 1:50,000 scale or up to phase level at 1:8000 or 1:4000 scale. For each soil mapping unit there will be association of soil taxonomic classes up to 1:12,500 scale as it is very difficult to establish a pure soil taxonomic class at small scale.

Finalisation of soil maps

The preliminarily interpreted maps with landscape units are finalized in light of ground truth collected and soil analytical data. The physiographic / landscape units are converted to soilscape units by incorporating the soil classification information into the mapping units. A final comprehensive legend showing the relationship physiography, lithology, relief, vegetation, land use along with description of soils and soil classification are developed and adopted for all soil mapping units in the study area. The final mapping units are transferred on to a suitable base map and final soil map is generated with appropriate legend.

The geo-pedological approach based on physiographic / landscape analysis work well upto 1:50,000 or 1:25,000 scale under most of the terrain situations. However, at larger than 1:25,000 scale this may not be suitable because the microrelief of locale specific conditions or soil phases play dominant role. The interpretation of satellite data is done by on-screen drawing of soil polygon boundaries using GIS software. This enables preparing soil digital database for further processing of soil information.

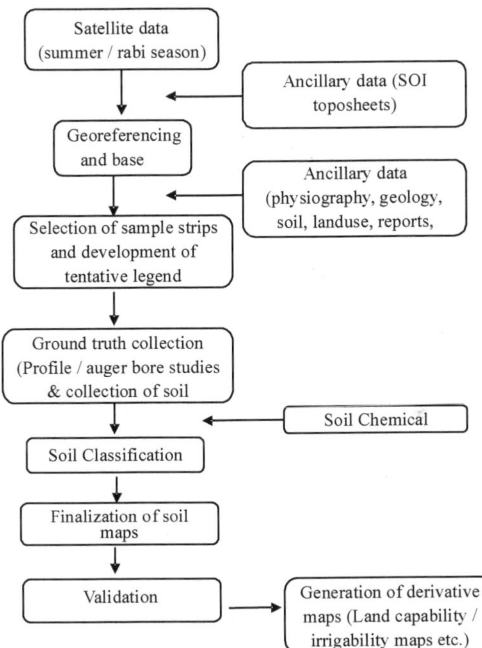

Fig. 2.2 Flow chart for Soil map preparation (*Source*: Remote sensing applications, NRSC, Hyderabad)

Digital techniques for mapping soils

New approaches like context classifiers, decision tree classififiers, neural network algorithms *etc.* are being employed during recent times to digital classification of soil resources. Digital image processing using supervised classification and unsupervised classification under maximum likelihood function are employed for soil mapping. In supervised classification, training sets (cluster of pixels with known composition after field work) are engaged in generation of soil maps. In case of unsupervised classification, cluster map showing the pixels with similar digital number (DN) is prepared. Field work to assess the soil composition is carried for each cluster. This ground truth information is then fed into the computer to generate soil maps.

3

Soil Classification

Soil Classification

Soil classification deals with the systematic categorization of soils based on distinguishing characteristics as well as criteria that dictate choices in use.

Soil classification is grouping of objects in orderly and logical manner into classes based on the properties of soils (differentiating characters) for the purpose of studying and identifying them. The soil individuals are grouped into classes of the lower category (soil series) which are grouped into classes of higher categories (soil orders). The lower categories are defined by large number of differentiating characters and higher category by a few differentiating characteristics.

For soil resources, experience has shown that a natural system approach to classification, i.e. grouping soils by their intrinsic property (soil morphology), behaviour, or genesis, results in classes that can be interpreted for many diverse uses. Differing concepts of pedogenesis and differences in the significance of morphological features to various land uses can affect the classification approach. Despite these differences, in a well-constructed system, classification criteria group similar concepts so that interpretations do not vary widely. This is in contrast to a technical system approach to soil classification, where soils are grouped according to their fitness for a specific use and their edaphic characteristics.

3.1 Soil Taxonomy: Purpose

Purpose of soil classification

- Organize knowledge leading to economy of thoughts
- Recognize properties of the soils
- Predict the behaviour of the soils
- Identify their potential uses
- Estimate their productivity
- To transfer agro technology from research farm to fields

3.2 Early and Modern Systems

Early Systems of Soil Classification

i) Economic classification

It is the grouping of soils based on their productivity for the purpose of taxation, adopted by the Revenue Department. The criteria used were soil colour and texture in combination with the potentialities for irrigation. Eg.. Soils suited for rice, groundnut and the system were of little importance and turned obsolete when the land use changed.

ii) Physical classification

It is grouping of soils based on their texture – a property closely associated with soil productivity and management. Eg. Loamy soil, sandy soil and clayey soil.

iii) Chemical classification

It is grouping of soils based on their chemical composition that have a bearing on their characteristics. Eg. Acidic, alkaline and calcareous soils. The system did not permit to classify all kinds of soils occurring in nature.

iv) Geological classification

It is grouping of soils based on the nature of underlying parent rock into two broad classes.

Residual or sedentary soils: The soils developed in-situ from the underlying rock, such as granite soils, soils from sandstone, limestone and basalt.

Transported soils: The soils developed on unconsolidated sediments such as alluvium colluvium or Aeolian.The system failed to recognize the influence of active soil forming factors like climate and vegetation on the parent materials. Eg. Granite rocks in the temperate region results in podzols whereas in tropical areas results in laterite soils.

v) Physiographic classification

It is grouping of soils based on the characteristics of landscape as terrace, mountain, valley, upland and lowland. The limitation of this system is that two or more soils may develop on the same landforms.

vi) Other systems

Based on organic matter – Mineral and organic soils

Based on soil structure – Single grained and aggregated soils.

Based on moisture or humidity, vegetation and temperature

–Arid soils, humid soils, sub humid soils, grassland soils and forest soils.

Recent Genetic Systems

i) Dokuchaev's Genetic System

Dockuchaev, the Founder of Modern Pedology and the Russian Scientist of 19th century divided soil into three categories as normal, transitional and abnormal soils based on soil genesis. These were later categorized as zonal, intrazonal and azonal soils.

Zonal soils: Soils with fully developed soil profiles and reflect the influence of climate and vegetation. Eg. Boreal – Tundra soils, Taiga – Light-grey podzolised soils, Steppe - Chernozem

Azonal soils: Soils with poorly developed soil profiles because of time as the limiting factor. Ex. Alluvial soils and Regosols

Intrazonal soils: Soils occurring within the zone and are largely influenced by the local conditions like topography, parent material *etc.,* Ex. Calcimorphic and Hydromorphic soils.

Limitation –Undue emphasis on climate and vegetation rather than on the intrinsic properties of soil.

ii) Coffey's concept

Coffee emphasized soils as independent natural bodies, classified on the basis of their own properties. He proposed five classes as arid soils, dark coloured Prairie sols, light coloured Timbered soils, Black Swamp soils and organic soils. Each class was subdivided into series on the basis of parent material and the series into types on the basis of surface soil texture.

iii) Marbut's Morpho Genetic System

Marbut of USA advocated classification of soil based on intrinsic soil properties. At the highest category level, he divided zonal soils into two classes as.

Pedalfers: Accumulation of iron and aluminium oxides, occuring in areas of high rainfall having surplus water for leaching.

Pedocals: Accumulation of calcium or calcium carbonate, occurring in areas of high evaporation having water deficit.

Limitation – The system was based on assumptions concerning soil genesis. Accordingly many soil series of USA cannot find a place in the system.

iv) Baldwin and Associate's Genetic System

Marbut's morpho genetic system was revised and elaborated by Baldwin, Kellog, Thorp and Smith.

A return to the zonality concept of Russian school Pedocal – pedalfer concept was deemphasized.

More emphasis was given on soil as a three dimensional body

The three soil orders zonal, intrazonal and zonal soils were divided into nine suborders on the basis of climate and vegetation. Each suborder was divided into great soil groups which were further sub divided into soil families, series and types.

Order	Suborder	Great soil groups
Zonal soils	Soils of the cold zone	Tundra soils
	Light coloured soils of arid regions	Sierozem soils, Desert soils
	Dark coloured soils of semi-arid, sub humid and humid grasslands	Prairie soils, Chestnut soils
	Soils of the forest grassland transition	Degraded Chernozem soils
	Light coloured Podzolic soils	podzolised soils
	Lateritic soils of warm-temperate and topical regions	Laterite soils
Intrazonal soils	Halomorphic soils	Solonchak or saline soils
	Hydromorphic soils	Bog soils
	Calcimorphic soils	Rendzina soils
Azonal soils	No suborder	Alluvial soils, Regosols

Limitations of Genetic Systems

- The two highest categories are defined in genetic terms and not on the basis of properties of soils.
- The definitions and concepts of the highest category *viz.,* order, interms of soil properties is not clear.
- The concepts and definitions of Great soil group are based on environmental factors rather than on soil properties.
- The properties of some soils were obvious under virgin soil conditions and were destroyed during cultivation and hence the classification of such arable soils became ambiguous.
- Possibilities to attempt definition of units in the lower categories based on soil properties for one interpretation doesnot hold good for the other interpretation.
- Nomenclature in the higher categories laid emphasis on colour or vegetation rather than on the salient properties of the soils.
- Nomenclature was evolved from several languages and it was difficult to name the intergrades.

Hence a desirable system of soil classification should be based on combinations of soil characteristics known to be significant to genesis and behavior, but not directly on the either.

To overcome the shortcomings of the earlier systems of soil classification, the U.S. Soil Survey Staff under the leadership of Guy D. Smith has developed a Comprehensive System of Soil Classification. Initially started in 1951, several approximations were made and a comprehensive system of soil classification, popularly called the 7th approximation was published in 1960 with supplements in 1964 and 1967. In 1975, the system was brought out as soil taxonomy (Soil Survey Staff, 1975).

3.3. Soil Taxonomy: A Comprehensive System

It is the basic system of soil classification for interpreting soil survey. It is a multi categoric system based on the concept of real bodies of soil.

Salient Features of Soil Taxonomy

The Comprehensive system is a morphogenetic system in which morphology of soil, that is an outcome of soil genesis, serves as a guide (Smith, 1963).

- Unlike the Genetic Systems, the Comprehensive System is based on measurable soil properties. Efforts have been made to define all classes in terms of soil properties that exist today
- It considers all such properties which affect soil genesis or are the outcome of soil genesis. Soil genesis forms the backbone of the comprehensive system. But it doesnot appear in the definition of the taxa.
- The common definition of a class of taxonomic system is type or orthotype.
- The nomenclature used in coining words is derived from Greek and Latin languages which is the most logical system.
- A new category i.e., sub group has been introduced to define the central concepts of great groups and their intergrades.
- Unlike the Genetic System, it is an orderly scheme without prejudices and facilitates easy remembering of the objects.

3.4. Diagnostic Horizons: Surface and Sub-surface

Diagnostic Horizons

These are horizons formed as result of pedogenic process. They have distinct properties that can be measured in terms of soil properties. They are not only useful in identifying soils but also in classifying soils at various category levels, especially great groups. There are two types of diagnostic horizons. These are surface (epipedon) and subsurface (endopedons) horizons.

Epipedons

An epipedon is the surface, or uppermost soil horizon. (Epi-over, pedon-soil) It may be thinner than the soil profile A horizon, or include the E or part or all of the B horizon. Epipedons derived from bedrock lack rock structure and are normally darkened by organic matter. Though the following seven epipedons are recognized, only three viz., mollic, ochric and umbric are of importance in the soils of India.

Mollic epipedon (L. Mollis – soft)

This epipedon is a soft dark grassland soil. A thick dark coloured mineral horizon with high (> 50%) base saturation and strong structure. It contains more than one 1 % organic matter with colour values darker than 5.5 dry and 3.5 moist. It is moist for more than nine months per year and cannot have both hard consistency and massive structure.

Anthropic epipedon (Gk. Anthropos – man)

While similar to the mollic epipedon, the anthropic epipedon contains greater than 250 ppm citric acid soluble P_2O_5 with or without a 50 percent base saturation and requires that the soil is moist three months or more over 8 to 10 years. It is commonly found in fields cultivated over long periods of time with additions of organic matter.

Umbric epipedon (L.Umbra – shade / dark)

Mollic-like in thickness, organic carbon content, color, P_2O_5 content, consistence, and structure, this epipedon has less than 50 percent base saturation (dominantly saturated with H^+). It is not naturally dry for more than three months a year.

Histic epipedon (Gk. Histos – tissue)

This organic horizon is water saturated resulting in reduced conditions to occur unless artificially drained. It is 20 to 60 cm thick and has a low bulk density often less than 1mg m^{-3}. The actual organic matter content is dependent on the percent clay. It is thinner than 30 cm, if drained or 45 cm if not drained.

Ochric epipedon (Gk. Ochros – pale, light coloured)

This horizon is light in colour and contains less than 1 % organic matter. It is dry for more than three months a year. The ochric epipedon extends to the first illuvial (B) horizon.

Plaggen epipedon (Gk. Plaggen – Sod)

This man-made horizon is 50 cm or more thick and has resulted from centuries of accumulation of sod, straw, and manure. It commonly contains artifacts such as pottery and bricks.

Melanic epipedon

A thick black horizon at or near soil surface which contain four to six percent organic carbon usually associated with short-range minerals or alluvium-humus complex. The deep dark colour is due to accumulation of organic matter resulting from root residues. Has a colour value (moist) and chroma of two or less throughout.

Follistic epipedon

Consists of organic soil materials that remain saturated for less than a month. The bulk density is less than 0.1 Mg /m3.

Grossarenic epipedon

A sandy (loamy fine sand or coarser) horizon, 100 cm or more thick over an argillic horizon.

Diagnostic subsurface horizons (Endopedons)

Diagnostic subsurface horizons (Endopedons) can be categorized as weakly developed horizons, as horizons featuring an accumulation of clay, organic matter, or inorganic salts, as cemented horizons, or as strongly acidic horizons.

Argillic (L. Argilla – White clay)

A silicate clay –enriched horizon formed by illuviation of clay. The fine clay is carried downward by percolating water and deposited as clay skins or cutans on ped faces on the walls of the pores. It is at least one-tenth the thickness of all overlying horizons. If the overlying horizon has less than 15 percent clay, the argillic has 3 percent more clay than the eluvial horizon above. If the overlying horizon has 15 to 40 percent clay, the argillic has 1.2 times that amount. If the overlying horizon has over 40 percent clay, the argillic has 8 percent more clay.

Natric (L. Natrium – sodium)

It is a high sodium clay enriched horizon with columnar or prismatic structure. The natric horizon is similar to the argillic horizon but has an exchangeable sodium percentage of 15 percent or more, SAR > 13, more exchangeable magnesium plus sodium than calcium plus exchange acidity (at pH 8.2).

Agric (L. Ager – field)

This horizon forms under a plough layer due to long and continued cultivation. It normally has lamellae (finger-shaped concentrations of material) of illuvial humus, silt, and clay.

Spodic (Gk. Spodos- Wood ash)

A humus and / or sesquioxides enriched horizon with or without iron. It is generally formed in cold humid regions. It must have 85 % or more of spodic materials in a layer of 2.5 cm. Such a horizon is rarely observed in India.

Sombric

A free draining horizon located not under an albic horizon. It has colours (darkness) and base status like an umbric epipedon. This has an illuvial accumulation of humus that is not associated with aluminum (spodic) or sodium (natric).

Cambic

This horizon (Structural B – horizon) shows some evidence of alterations due to physical movement and chemical weathering but is very weakly developed between A and C horizons. The cambic horizon has less illuviation than found in the argillic and spodic horizons. The pedogenic processes have altered the material to form structure to liberate free iron oxides to form silicate clays or both. The horizon is extremely variable in mineralogy because of its pedogenic youthfulness, occurs under differing environments and may develop in the presence or absence of fluctuating ground water.

Kandic

The kandic is a horizon with an illuvial accumulation of 1:1 (kaolinite-like) clay with or without clay skins and has a CEC of less than 16 cmol kg^{-1} clay at pH 7 and an ECEC of less than < 12 cmol kg^{-1} clay. A clay increase within 15 cm of the overlying horizon is 4 percent or more if the surface has less than 20 percent clay; The horizon is 15 -30 cm thick. Organic carbon constantly decreases with increasing depth.

Oxic (Oxide minerals)

A horizon enriched with Fe and Al-oxides with dominance of 1:1 type clay minerals and from which silica has leached. It is atleast 30 cm thick and is sandy loam of finer in texture. The oxic horizon contains highly weathered clays. It has a CEC of less than 16 cmol kg^{-1} clay pH 7 and an ECEC of less than < 12 cmol kg^{-1} clay. The clay content increase is gradual than in Kandic horizon. It contains less than 10 % weatherable minerals in the sand fraction.

Sulfuric

The sulfuric horizon forms as a result of draining soil with a high sulfide content that is oxidized to sulfates, drastically reducing the pH. It is at least 15 cm thick and has a pH of 3.5 or less, which is toxic to plant roots, and has yellow mottles of jarosite.

Salic

Measuring 15 cm or more thick, the salic horizon contains at least 2 percent water soluble salts like NaCl, Na_2SO_4. A 1:1 soil to water extract has an electrical conductivity of 30 dS/m (decisiemens per meter) or more.

Albic (L. Albus – White)

A bleached (Clay, humus, and other coatings have been leached from this eluvial horizon, leaving light colored sand and silt particles) E horizon of planosols and podzols. It has typical colour values of > 5 (dry) or > 4 (moist).

Glossic

This transitional horizon has parts of an eluvial horizon and the remnants of a degrading argillic, kandic, or natric horizon. It is 5 cm or more in thickness.

Calcic (L. Calcic - Lime)

A horizon with secondary Ca and / Mg- carbonate enriched materials. It is 15 cm or more thick, and has evidence of calcium carbonate movement. If the horizon is cemented, it is classified as petrocalcic. It has 15% or more of secondary accumulation of carbonate and contains atleast 5% more carbonate than any underlying horizon.

Gypsic

Calcium and or magnesium sulphate – enriched horizon. It is more than 15 cm thick and contains atleast 5 % more $CaSO_4$ than the underlying horizon.

Hardpan horizons

Petrocalcic

An indurated calcic horizon that has hardness of 3 or more (Moh scale) and whose one- half or more of the dry fragments break down in acid, but not in water. It is cemented by carbonates and not penetrable by spade or auger.

Petrogypsic

A strongly cemented gypsic horizon whose dry fragments donot slake in water. Cementation restricts the penetration of roots.

Placic

A thin (2 to 10 mm thick), slowly permeable, dark reddish brown to black coloured cemented by iron and manganese that lies within 50 cm of the surface.

Duripan (L.durus – hard, pan – hard pan)

This subsurface horizon is cemented by silica in more than 50 percent of its volume. It dissolves in concentrated basic solution or alternating acid and then basic solutions, but does not slake in 18 % HCl.

Fragipan (L.fagillus – brittle, pan- hard pan)

A fragipan is a brittle horizon situated at some depth below an eluvial horizon. It has a low organic matter content, lower bulk density than overlying horizons, and hard or very hard consistence when dry. It is mottled, slowly or very slowly permeable to water and shows bleached cracks forming polygons. Base saturation and pH are normally low.

Plinthite (Gk. Plinthos - brick)

Plinthite is iron-rich, humus –poor mixture of clay with quartz and other diluents. It occurs as dark red mottles, which are in platy, polygonal and reticular patterns. Plinthite changes irreversibly to an ironstone hardpan, especially if it is exposed to heat from the sun.

Diagnostic organic soil materials

There are three kinds of diagnostic organic soil materials based on the degree of decomposition of the plant materials. They are (i) fibric (ii) Hemic (iii) Sapric.

Fibric soil material (formerly peat)

The fibrous material in an unrubbed condition, constituting over 2/3 of the mass and yields almost clear solution when extracted with sodium pyrophosphate.

Hemic soil material

The fibrous material in an unrubbed condition, constituting 1/3 to 2/3 of the mass and in an intermediate stage of decomposition.

Sapric soil material (Gk. Sapros – rotten, formerly muck)

The most highly decomposed kind of material that has identifiable fibrous material, in an unrubbed condition, constituting less than 1/3 of the mass.

Hemi illuvic material

Illuvial humus that accumulates after prolonged cultivation of some acid organic soils.

Limnic material

Organic or inorganic materials deposited in water by precipitation or through the action of aquatic organisms such as algae, or derived from underwater ad floating aquatic plants modified by micro organisms. E.g. Marl, diatomaceous earth and sedimentary peat.

Soil Moisture Regime

The term "soil moisture regime" refers to the presence or absence either of groundwater or of water held at a tension of less than 1500 kpa in the soil or in specific horizons during periods of year. Water held at a tension of 1500 kpa or above is not available to keep most mesophytic plants alive. Five moisture regimes have been recognized.

1. ***Aquic*** (L. aqua, water): It refers to a reducing regime devoid of dissolved oxygen as it is saturated by ground water itself or capillary rise.
2. ***Aridic and torric*** (L. aridus, dry, and L. torridus, hot and dry): In the aridic moisture regime, the moisture control section (soil depth varying from 10 to 60 cm depending upon soil texture) in most years is i) dry in all parts for more than half of the cumulative days per year when the soil temperature at adepth of 50 cm from the soil surface is above 5° C and ii) moist in some or all parts for less than 90 consecutive days when the soil temperature at a depth of 50 cm is above 8° C. such a situation normally occurs in arid climate.
3. ***Udic*** (L. Udus, humid): In most years soil moisture control section is not dry in any part for as long as 90 cumulative days
4. ***Ustic*** (L. Ustus, burnt, dryness): This regime is intermediate between the aridic and the udic regime. Its concept is one of moisture that is limited but is present at a time when conditions are suitable for plant growth.
5. ***Xeric*** (Gr. Xeros, dry): This regime is typified in the Mediterranean climates where winters are moist and cool and summers are warm and dry. In areas of xeric moisture regime, the soil moisture control section in normal years, is

dry in all parts for 45 or more consecutive days in the 4 monts following the summer solstice and moist in all parts for 45 or more consecutive days in the 4 months following the summer solstice and moist in all parts for 45 or more consecutive days in the 4 months following the winter solstice.

Soil Temperature Regime

Soil Temperature Regime constitutes one of the two principal elements of the pedoclimate (along with soil moisture) its variations exercise a major effect, on the behaviour of plants on one hand, and or pedogenesis on the other. In soil taxonomy, Soil Temperature Regimeare considered in definition of the taxonomic units and are defined by temperatures taken in a soil at 50 cm depth or at a lithic or paralithic contact situated at less than 50 cm.

The following soil temperature regimes are used in defining classes at various categoric levels.

- ***Cryic***: The soil has a mean annual temperature higher than 0° C, but lower than 8° C but do not have permafrost.
- ***Frigid***: A soil with a frigid temperature regime is warmer in summer than a soil with a crying regime, but its mean annual temperature is lower than 8° C and the difference between mean summer (June, July, August) and mean winter (December, January and February). Soil temperatures is more than 6° C either at a depth of 50 cm from the soil surface or at a dense, lithic or paralithic contact, whichever is shallower.
- ***Mesic***: The mean annual soil temperature is 8° C or higher but lower than 15° C and the difference between mean summer and mean winter soil temperatures is more than 6° C either at a depth of 50 cm from the soil surface or at a densic, lithic or paralithic contact, whichever is shallower.
- ***Thermic***: The mean annual soil temperature is 15° C or higher but lower than 22° C and the difference between mean summer and mean winter soil temperatures is more than 6° C either at a depth of 50 cm from the soil surface or at a densic, lithic or paralithic contact, whichever is shallower.
- ***Hyperthermic***: The mean annual soil temperature is 22° C or higher and the difference between mean summer and mean winter soil temperatures is more than 6° C either at a depth of 50 cm from the soil surface or at a densic, lithic or paralithic contact, whichever is shallower.

If the name of a soil temperature regime has the prefix iso, the mean summer and winter soil temperatures differ by less than 6° C at a depth of 50 cm or at a densic, lithic or paralithic contact, whichever is shallower.

- ***Isofrigid***– The mean annual soil temperature is lower than 8° C
- ***Isomesic***- The mean annual soil temperature is 8° C or higher but lower than 15 ° C
- ***Isothermic***- The mean annual soil temperature is 15° C or higher but lower than 22° C
- ***Isohyperthermic***- The mean annual soil temperature is 22° C or higher

Family and series differentiae and names

The following differentiae are used to distinguish families of mineral soils and the mineral layers of some organic soils within a subgroup. The class names are used to form the family name.

Particle size classes

- Sandy skeletal
- Loamy skeletal
- Clayey skeletal
- Sandy
- Loamy
- Coarse loamy
- Fine loamy
- Coarse-silty
- Fine silty
- Clayey
- Fine
- Very fine

Mineralogy classes

Gibbsitic, ferruginous, kaolinitic, halloysitic, gypsic, carbonatic, illitic, vermiculitic

CEC classes (CEC of fine earth fraction)

Super active = 0.6or more

Active = 0.4 to 0.6

Semiactive = 0.24 to 0.4

Subactive < 0.24

Soil temperature classes

Frigid <8° C

Mesic 8° C to 15° C

Thermic 15° C to 22° C

Hyperthermic 22° C or higher

Isofrigid <8° C

Isomesic 8° C to 15° C

Isothermic 15° C to 22° C

Isohyperthermic 22° C or higher

3.5. Soil Taxonomy

Definition of Soil Taxa

In the definition of soil taxa, differentiating characteristics selected are properties of soils, including soil temperature and moisture regimes. The genesis of soil is neither directly deployed nor is in the forefront, except as a guide to relevance and weighing of soil properties.

Nomenclature

The basic principles followed in coining names, according to Heller (1963) are that the name

- Should be most easily remembered
- Should suggest some properties of the object
- Should suggest the place of a taxon in the system
- Should be as short as possible
- Should be as euphonic as possible and
- Should fit readily in as many languages as possible.

Criticism of Soil Taxonomy

- The recently developed system of soil classification viz., soil taxonomy (1975) apparently departs from the genetic approach
- The system does not have strong geographic bias towards the four orders viz., Entisol, Vertisol, Inceptisol and Histosol
- The soils with a different genesis but with identical properties are classified within the same unit

- There is no particular order for the strictly hydromorphic and saline-sodic soils, as in the case of other systems.

Appreciation of Soil Taxonomy

- It is the most elaborate system marked by great care and precision
- The primary basis for identifying different classes in the system are properties of soils as they exist in the field
- The nomenclature (with Latin and Greek origin) gives a definite composition of the major soil characteristics
- The system has in-built mechanism to permit addition of new soil groups. E.g. the new orders Andisols and Gellisols have been included in the system recently
- It permits classification of soils rather than soil forming process
- It permits the classification of soils of unknown genesis.

Structure of soil taxonomy

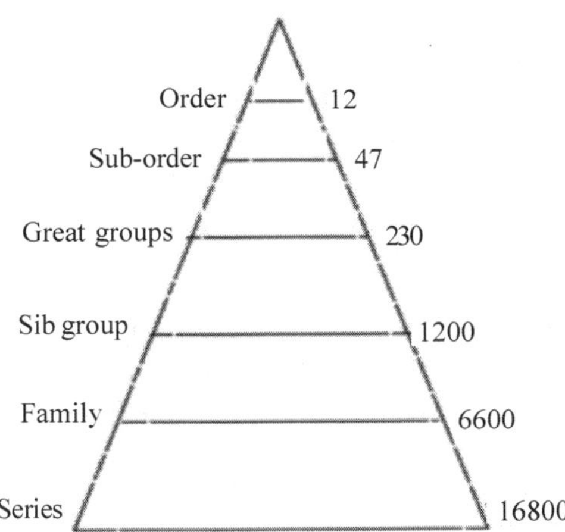

Higher categories

1. **Order:** The number of taxon is 12 (AVAGAMIHOUSE). Aridisol, VErtisol, Alfisol, Andisol, Mollisol, Inceptisol, Histosol, Oxisol, Ultisol, Spodosol, Entisol and Gellisol. Differentiating character is based on morphology as indicated by the presence and absence of diagnostic horizons.
2. **Suborder:** It is the sub division of the order based on genetic homogeneity, wetness, climatic environment, parent material and vegetation effects. There are 47 suborders

3. **Great groups:** The major emphasis is on the diagnostic horizons and presence or absence of diagnostic layers (Plinthite, fragipan, duripan), base status, soil temperature and moisture regimes They are 230 in number. They are based on similarities in soil moisture, temperature and base status. It also involves the diagnostic sub surface horizons.

Lower categories

4. **Sub groups :** The typic is used to define the central concept of a great group;the orders are used to indicate integrades to other great groups, sub orders and orders and the extragrade to not soil. A new category designed to define the central concept of a great group. They are 1200 in number. Three types of subgroups are typic, intergrade and extragrade.
5. **Family :** Number of families in soil is 6600. Families are based on texture, mineralogy, depth, temperature and calcareousness. It is a practical category for making predictions and land use plans. This is a group of soil series within a sub group that owe their parentage to the similar kind of parent material and that are relatively homogenous with respect to soil air, soil water and plant root relationships.
6. **Series :** Soil series is considered as the most important category in the system as it is the fundamental unit of classification and is also the basic unit for most soil mapping. A group of soils having soil horizons, similar in differentiating characteristics and arrangements within in the soil profile, except for the texture of the surface soil, and developed from a particular type of parent material"

The lowest and the most specific category. The series is a collection of soil individuals, essentially uniform in differentiating characteristics (like colour, texture, structure, consistency, pH and EC) and its arrangement of horizons. The series are named after the geographic name of the place where it was first recognized.

7. **Soil Types:** It is a subdivision of a Series based on the texture of the surface soil. Many soil series, especially those derived from homogenous parent materials have only a few types.
8. **Soil phase :** This is a subdivision of any class in the taxonomic system but not itself a category of the system. Features such as variation in slope, degree of erosion, stoniness, and thickness of alluvial deposits have customarily been used to separate phases within a soil type or soil series. Based on differences in surface soil texture, solum thickness, percentage slope, the stoniness, the saltiness, the extent of erosion damage and other conditions.

S.No	Order	Formative Element	Derivation of formative element	Soils belonging to the order
1.	Entisol	Ent	Recent	Azonal soils, Gley soils
2.	Vertisol	Ert	L. Verto, turn	Grumusols, Black cotton soils with swelling and shrinkage clay
3.	Inceptisol	Ept	L.inception, beginning	Brown forest soils
4.	Aridisol	Id	L. aridus, dry	Desert soils of Rajasthan, solonchak, solonetz
5.	Mollisol	Oll	L. mollis, soft	Chernozem soils, Chestnut soils
6.	Spodosol	Od	Gk. Spodos, wood ash	Podzols (Ashy soils)
7.	Alfisol	Alf	Non sense symbol	Red soil, Planosols, Half–Bog soils, Non – calcic brown soil
8.	Ultisol	Ult	L. Ultimus, last	Lateritic soils
9.	Oxisol	Ox	Oxide soil	Laterite soils
10.	Histosol	Hist	Tissue soil	Peat soil , Bog soils (> 30 % M)
11.	Andisol	And	Modified from Ando	Volcanic ash soils
12.	Gelisol	El	Frost churning or mixing	Frozen Tundra soils exhibiting Cryoturbation

3.6. Soil Orders

- In Soil Taxonomy, there are twelve orders. The names of twelve orders can best be remembered from the phrase A Vagami House each word of which denotes an order name. For instance a- Alfisols, Andisols or Aridisols; v-Vertisols, h-Histosols etc. the diagnostic properties of all the twelve soil Orders are briefly

Table 3.1: Diagnostic properties of different soil orders.

Based on	Diagnostic property	Order
Nature of soil material:	-Very high content of organic matter	*Histosols*
	-cracking or swell - shrink clays	*Vertisols*
	-Black, low bulk density volcanic (andic) material	*Andisols*
Presence/absence of diagnostic horizon	Spodic horizon	*Spodosols*
	Oxic horizon	*Oxisols*
	Argillic horizon with low base saturation	*Ultisols*
	Argillic horizon with high base saturation	*Alfisols*
	Mollic epipedon and high base saturation	*Mollisols*
	Cambic horizon, an umbric or plaggen epipedon etc.	*Inceptisols*
	No distinctive horizon	*Entisols*
Soil moisture regime	Aridic or dry, or saline conditions	Aridisols
Frost Churning	Cryoturbation	Gelisols

1. Entisols (Recent)

Concept

- The Entisols are very recently developed
- Mineral soils with no diagnostic horizon other than an ochric or anthropic epipedon
- Slight degree of profile development due to lack of time to form horizon, extreme texture, exceedingly unfavourable climatic or topographic conditions.
- The parent materials may be recent alluvium, sand dunes or variety of rocks.

Distribution

- Entisols are widely distributed in steep slopes, flood plains, near the oceans and in sand dunes irrespective of climate, vegetation and topography.
- In India these are common in rocky humid or subhumid mountain regions and in sand bars of Rajasthan, Haryana and Central Punjab.

Major suborders

Aquents, Fluvents, Psamments, Arrents and Orthents

Land Use

- Coarse textured with poor water and nutrient holding capacities.
- Physical conditions are poor especially soil structure and is susceptible to wind erosion.

- Short duration crops, grains, vegetables, pastures can be grown on conserved moisture.

2. Inceptisols (L. Inceptum, Beginning)

Concept

- They represent early stage of soil formation.
- They have accumulation of clay in their subsurface, but it is insufficient to qualify it as an argillic horizon, which is diagnostic for Alfisols and Ultisols.
- They have dark coloured surface horizon which is limiting in depth, organic matter and base saturation to qualify it as a mollic epipedon (Mollisols).
- These soils are usually not dry and have one or more of the diagnostic horizons (cambic, umbric or mollic with low base saturation).
- They have developed rather recently owing to the alteration of the parent material but without much leaching and accumulation of materials in sub soil.
- The soils have too weak profile development to be called zonal yet have horizonation enough not to be termed as Azonal.
- The order includes many agriculturally productive soils.
- It also includes others whose productivity is limited because of imperfect drainage and sodicity.
- The soils develop under varied climatic and vegetation conditions.

Distribution

These are widely distributed the world. They occur in most climatic regimes except the arid regions such as Thar desert in Western Rajasthan. They are common in inter tropical regions under all topographical positions.

Major suborders

Aquepts, Anthrepts, Cryepts, Ustepts, Xerepts and Udepts.

Land Use

- Limiting soil temperature (cryic) and moisture (ustic and xeric regimes permit farmers to have one crop a year.
- Petrocalcic horizons pose physical limitations and sulphuric acid poses chemical limitations.

- These are agriculturally productive and provide natural grazing grounds. Pastures, vegetables and grain crops are commonly cultivated.

3. Vertisols (L.Verto, Turn or Churn)

Concept

- These are uniform, thick (atleast 50 cm) tropical, black and dark coloured, cracking-clay, mineral soils that have high clay content (> 30 %).
- These soils swell on wetting and shrink on drying. The swell – shrink process induces the development of wide, deep cracks associated with gilgai microrelief or intersecting slickensides (Polished ped faces in the subsoil).
- They develop on basic parent materials such as basalt, limestone, shale or marls of smectitic composition. They occupy lower topographical position or flat terrain.
- High clay content and swell and shrink nature of clay is essential for the formation of vertisols.

Distribution

- They are found in sub humid to semi arid climates where annual precipitation varies from 500 to 1500 mm.
- In India they are observed in Madhya Pradesh, Maharashtra, Gujarat, Andhra Pradesh, Tamil Nadu and Rajasthan.
- In India these are termed as Black cotton soils (Regur).

Major suborders

Aquerts, Cryerts, Xererts, Torrerts, Usterts, Uderts.

Land Use

- Because of shrink-swell characters, they pose many tillage problems during cultivation, or when used for building foundations, laying drainage and gas pipelines, installing irrigation networks etc., Tilting of trees, fences, poles and sinking of floors are characteristics of this order.
- Grasses, cotton, sorghum, pigeonpea, paddy etc., may be cultivated. They are very productive, if managed properly.

4. Mollisols (L.Mollis, Soft)

Concept

- These are soils of grassland vegetation under semi arid to humid environment.

- They have a dark coloured, base-rich, well-structured (granular or crumb) surface horizon, that is rich in organic matter called the mollic horizon.
- The subsurface horizons, rich in illuviated clay (argillic), calcareous (calcic) or gypsiferous (gyspic) materials may be present.
- Roots of grassy vegetation decompose resulting in dark stable compounds associated with soil burrowing activity of diverse populations of soil invertebrates, especially earthworms.
- They develop under a wide range of temperatures from the equator to the poles and in lowlands to high mountain meadows.
- They are formed in parent materials rich in bases such as limestone, basalt, marl or alluvium. They have high base saturation and abundant calcium.

Distribution

- This order is observed in the central USA, Russia and in the central South America.
- In India, such soils are observed in the 'Tarai' region of Uttar Pradesh and Uttaranchal, in Himalayas of Himachal Pradesh, northern Bihar and in Maharashtra and Madhya Pradesh in association with vertisols.

Major suborders

Albolls, Aquolls, Rendolls, Cryolls, Xerolls, Ustolls and Udolls.

Land Use

- The mollisols are inherently the best agricultural soils without real constraints.
- In India, these soils produce optimum yields both under irrigated and unirrigated conditions with minimum inputs. Maize, sorghum, field and vegetable crops are cultivated in this soil.

5. Aridisols (L.Aridus, Dry)

Concept

- These are mineral soils of arid and semi-arid regions and of areas of high groundwater table.
- The soils remain dry for most part of the year and salts accumulate at the surface and / or in the solum resulting in the development of a salic, gypsic or calcic horizon.

Distribution

- They occur widely in the arid climatic environments of the world like the Sahara, Gobi and Thar deserts and occupy over a billion hectares of land area.
- They are dominantly observed in S.Asia, N and SE Africa, Australia, South America, SW and N.USA, S.Africa and Russia. In India they occur in Rajasthan, Gujarat, Haryana and in Leh and Ladak regions.
- The important soil forming processes responsible for the evolution of such soils are salinization and calcification. If the groundwater is brackish, salt accumulation takes place in the profile by capillary fringe.
- The parent material is alluvium, loess or till.

Major suborders

Cryids, Salids, Durids, Gypsids, Argids, Calcids and Cambids

Land Use

- The vegetation in this order is sparse and xerophytic.
- They have limitations of aridic climatic conditions and brackish groundwater close to the surface.
- The sub surface horizons hamper root growth and water movement.
- If the gypsiferous landscapes are irrigated, they develop sink holes due to dissolution of gypsum and the land sinks behind leaving uneven surfaces for cultivation.
- These soils require irrigation for crop growth. They are best suited for sparse native grazing.

6. Alfisols (Pedalfer of Marbut)

Concept

- Alfisols are base – rich, timbered, mineral soils of sub humid and humid regions.
- They have light coloured ochric epipedon, over a subsurface argillic horizon that is rich in exchangeable cations with base saturation of > 35%.
- Alfisols are more strongly weathered than the inceptisols, but less than the Ultisols.
- The Alfisols develop under semi-arid to humid climates of temperate and tropical climates and where the vegetative cover is from grassland to forest.

- Though the soils are formed in varying topographic conditions, steep slopes are not conducive because of surface run off. Decalcification, eluviations and illuviation results in the formation of argillic horizon.
- In sodic soils, high sodium is responsible for the dispersion and mobilization of clay.

Distribution

- In India, red soils are found in Andhra Pradesh, Assam, Bihar, Arunachal Pradesh, Himachal Pradesh, Maharashtra, Karnataka, Tamil Nadu, Orissa and Madhya Pradesh.
- They are found in localized depression in soils having high sodium on the exchange complex.

Major suborders

Aqualfs, Cryalfs, Ustalfs, Xeralfs, Udalfs

Land Use

- Alfisols are naturally fertile and productive. They are used for forestry, grazing and for growing fruits, vegetables and grain crops.
- The sodium-rich Alfisols need ameliorative measures such as gypsum application before bringing under cultivation.

7. Spodosols (Gk. Spodos – wood ash)

Concept

- These are mineral soils with accumulation of sesquioxides and humus in the sub surface horizons.
- The diagnostic features of these soils is an illuvial horizon, enriched with free sesquioxides and humus underlying a belached, wood-ash-coloured eluvial E-horizon.
- In the past, these soils were classified as podzols.
- The soils develop on coarse-textured (sandy) parent material supporting acid producing vegetation like coniferous under cold and temperature climate conditions.
- Podzolisation is responsible for the development of spodosols. There is limited polysaccharide production due to reduced bacterial activity. Organic acids result in the formation of organic-sesquioxide-clay complex, which are soluble thus forming a spodic horizon.

- Spodosols also develop due to degradation of Alfisols.

Distribution

These soils are widely distributed NW Europe, NE USA, Canada etc., In India, true Spodosols are seldom reported to occur because of the lack of siliceous parent material and environmental conditions (cool humid, supporting acid producing vegetation) needed for the formation.

Major suborders

Aquods, Cryods, Humods, Orthods

Land Use

Spodosols are naturally unfertile and are used for forestry, soft-wooded timber production and pasture land. When fertilized they may support one crop in a year because of the limiting climatic condition.

8. Ultisols (L.Ultimum – Last)

Concept

- The Ultisols are comparable with Alfisols, but have low base saturation (< 35 %) due to advanced stage of weathering.
- These are base poor, timbered mineral soils of humid regions developed under high rainfall and forest vegetation.
- They have argillic or kandic subsurface horizon.
- These soils are completely decalcified by the leaching of bases during long period (thousands of years) of their pedogenic evolution.
- Base poor clay minerals such as kaolinite, predominates whilst gibbsite is occasionally observed.

Distribution

- The soils are most extensive in the warm, humid and (subtropical) regions of the world. These are most predominant in SE USA, central Africa and SE Asia.
- In India, they are observed in Kerala, Tamil Nadu, Karnataka, Orissa, Assam and north-eastern states.

Major suborders

Aqults, Humults, Udults, Xerults

Land Use

- Because of the low fertility and low base status, these soils pose limitations for agricultural use. Generally, these soils are used for forestry, but may produce good agricultural, vegetable and plantations crops, when they are adequately limed and fertilized.
- In the tropical regions, such soils are cultivated for pineapple, sugarcane, coffee, cocoa, coconut, rubber etc.,

9. Oxisols (Oxide - Iron)

Concept

- These are extremely and deeply weathered mineral soils of the humid tropics that are very poor in soil fertility.
- These are characterized by a uniform profile having limited or negligible amount of weatherable minerals, and are dominated by kaolinitic and sesquioxide – rich deep subsurface diagnostic horizon (oxic).
- Deeply weathered, mottled horizon (plinthite) may also develop.
- Most oxisols have brick-red colour, but some are also yellow or grey in colour.
- They develop from basic parent materials with sufficient iron-bearing ferro – magnesian minerals (pyroxene, amphibole, biotite). Silica is removed leaving behind a high proportion of the oxides of iron and aluminium.
- The hydrated form of sesquioxides, being unstable, get dehydrated irreversibly to Al_2O_3 / Fe_2O_3 due to high temperature. Such a material is cut into bricks and used in construction of buildings.
- These soils are rich in clay content, but unlike Vertisols, the clay in Oxisols is non-sticky and predominates in kaolinite and gibbsite minerals, that have very low exchange capacity.

Distribution

- Oxisols are observed predominantly in the tropical world, on gently sloping very old landscapes in Central Africa, South America and South-east Asia.
- In India, such soils are found in Kerala, Tamil Nadu, Karnataka and Orissa. Their occurrence in other areas is indicative of climatic change in the recent past.

Major suborders

Aquox, Torrox, Ustox, Perox, Udox

Land Use

- Oxisols are chemically degraded soils and need careful management and fertilization for agricultural use.
- They pose the problem of phosphate fixation as Fe and Al. These are mainly used for grazing and forestry.
- Some are used for growing crops such as coffee, rubber, cocoa, sugarcane and tropical fruit crops (pine apple, coconut, jackfruit etc.,).
- Shifting agriculture is a common practice adopted in such areas. The cycle is reduced to 3 o 5 years, resulting in serious problems of soil erosion by water and nutrient depletion.

10. Histosol (Gk. Histos - Tissue)

Concept

- These are organic rich soils with high, peaty horizon (histic epipedon) that develop in lower topographic positions in a permanent water saturated environment.
- As a thumb rule, a soil without permafrost is classified as a Histosol if half or more the upper 80 cm is organic.
- The main process in their formation is accumulation of peat.
- The breakdown of organic matter is related to waterlogging as the most effective microbial decomposers require oxygen. Sediment or rock (R,C) is little altered by weathering.
- There may be some leaching or formation of gley minerals such as pyrite or siderite, but most of the weatherable minerals and structures of the parent materials are retained.
- Histosols support bog, swamp and marsh. The organic matter content ranges from 20 to 30 percent.
- In the past, these soils were known as peat, muck and bog soils.

Distribution

- Histosols occur in selected parts of Europe and Asia.
- In India, Histosols have not been reported, if they exist, these may be of rare occurrence.
- They are found to occur in Kerala and Andaman and Nicobar islands.

Major suborders

Hemists, Folists, Fibrists, Saprists.

Land Use

- These soils are used for various forms of extensive forestry and /or grazing
- In Kerala, they are used for growing paddy, soon after draining the flood / run off water received during the monsoon period, but keeping the subsurface soils submerged in water to avoid oxidation of pyrite to form sulphuric acid.
- Under natural conditions, they support mangrove vegetation that helps in maintaining ecobalance in coastal regions of Orissa, Tamil Nadu, Goa and Sunderban areas of West Bengal.

11. Andisol (Japan, Ando – Black soil)

Concept

- Andosols were first recognized in Japan and were named as Volcanic ash material.
- They are dark coloured, low bulk density soil that have developed on volcanic ash parent material.
- They donot have an albic horizon but have andic properties : low bulk density (< 0.9 Mg m^{-3}) or 60 percent or more vitric volcanic ash within 60 cm of the mineral soil surface.
- It has high allophane which gives them a very low bulk density (easy to cultivate) and fluffiness, especially in the B horizon. They are generally very fertile.

Distribution

- Andisols are observed in all topographic positions and under all soil moisture and temperature regimes.
- They are found in the (sub)tropical and Meditterranean climates, but not in cold, temperate conditions where the soils have humus rich surface horizon.
- Since they are formed from volcanic ash or materials derived from it as volcanic alluvium, volcanic loess, they are dominantly observed in the 'Ring of Fix' in the Pacific, including Japan, New Zealand, Indonesia and Phillipines.
- They are also observed in the Rift Valley of Africa, especially Kenya, western coast of USA and South America.

- In India, such soils have not been reported to occur so far, although volcanic activity is locally observed in Andaman and Nicobar islands, where there is possibility of localized occurrence.

Sub orders

Aquands, Cryands, Torrands, Xerands, Vitrands, Ustands and Udands.

Land Use

- Andisols that occur on steep slopes, have the limitation of mechanized farming.
- The major limitation is their high phosphate fixation capacity.
- Andisols are stable and resist water erosion because of high infiltration and permeability rates.
- When dry, they are susceptible to wind erosion. They demand careful management for engineering use because of low bulk density, poor compactability and large changes in cohesion on drying.
- It is advantageous to grow crops that show little response to the application of phosphorus. E.g. Sweet potatoes.

12. Gelisol (Gelic - Churning)

Concept

- These are the soils with gelic materials viz., mineral or organic soil materials that show evidence of cryoturbation(frost churning) underlain by permafrost within 2 m of the soil surface.
- Diagnostic horizons may or may not be present in Gelisols as thawing and freezing play an important role in evolution.
- Permafrost acts as a barrier to the downward movement of the soil solution.
- Cryoturbation (frost churning or mixing) is an important process in the development of Gelisols.

Distribution

- Gelisols are observed in areas of permafrost, that include extreme northern USA, Canada, Greenland, Russia, Mongolia and other areas in the extreme north of the northern hemisphere.
- In India, such soils have not been observed and reported.
- They may be observed in permafrost conditions (ice caps in the extreme North, represented by Leh and Ladakh and Sikkim regions of the Higher Himalayas)

Sub orders

Histels, Tubels and Orthels.

Land Use

- Gelisols are not cultivated because of the limitation of severe cold climate. They are left to support natural vegetation. Orthels of anhydrous condition, support little vegetation.
- Entisols are the least developed and Oxisols represent the most advanced stage of weathering in evolution of soils.

3.7. Soils of India

India has a geographical area of 327.4 million ha (3274690 km^2) and lies between latitudes of 8°C and 37°C and longitude of 69° and 93°E. The nature and properties of soils are greatly dependent upon the soil forming factors and processes. Since there exists significant differences in these parameters, these are expected to exert their influence in the evolution of soils of India. India represents all the major soil groups of the world.

Physiography

- India has been divided into three broad regions *viz.,*
- Traingular plateau of the peninsula (in the Deccan and South of the Vindhyas)
- Mountain Region of the Himalayas which borders India from the west, North and East and is known as Extra –peninsula
- Indo-Gangetic Plain of Punjab and Bengal, separating the two above mentioned areas and extending from valley of the Idus from the sind to that of the Brahmaputra in Assam.

Geology

The greater part of the peninsula is occupied byArchaen rocks, comprising gneisses and schists and other igneous and metamorphic rocks, Cuddapah and Vindhyan rocks followed by coal bearing Gondwana formations. These are mainly distributed over the north-central and NE-central parts where Red or Yellowish-Red (Alfisols, Inceptisols, Entisols) have developed. The western and central zones of India are covered by lava flows of the Deccan Trap where basaltic rocks predominate where black soils have formed (Vertisols, Inceptisols).

The extra peninsula, shows the development of marine sediments in the North of Himalayas. The major rock formations are old sedimentary (sandstone, limestone) and igneous (granite) that at places are metamorphosed to gneiss and schists (Alfisols, Inceptisols).

The vast Indo-Gangetic and other plains are composed of alluvia of the great river system flowing in this region. The soils belong to the orders of Entisols, Inceptisols, Alfisols and Mollisols.

Soils

India with a great variety of landforms, geological formations and climatic conditions exhibit a large variet of soils. Schokalskaya (1932) prepared the first soil map of India and classified the soils on the basis of existing knowledge of Russian pedological principles. Govindarajan (1965) divided the soils of India into 24 major soil groups. Raychaudhuri and Govindarajan (1971) revised the soil map of India giving the equivalents of different soil groups in US comprehensive system of soil classification. The Nation Bureau of soil survey and Land use planning (1985) prepared a soil map using the units of soil taxonomy.

S. No	Major Soils (Generic System)	Area (M ha)	Equivalent soil names	
			US Soil Taxonomy	FAO Legend
1.	Alluvial soil	75	Inceptisols, Entisols, Alfisols, Aridisols	Cambisols, Fluvisols, Luvisols, Gleysols, Solonchalks, Solonetz
2.	Black cotton soils	72	Vertisols, Inceptisols, Entisols	Vertisols, Cambisols, Leptisols, Regosols
3.	Red soils	70	Alfisols, Inceptisols, Entisols	Luvisols, Cambisols, Leptosols, Regosols
4.	Laterite and lateritic soils	25	Oxisols, Ultisols, Inceptisols	Cambisols, Alisols, Leptosols, Plinthosols
5.	Saline and Alkali soils	7.5	Aridisols, Inceptisols, Alfisols, Entisols, Vertisols	Solonetchaks, Solonetz, Cambisols, Luvisols, Arenosols, Vertisols
6.	Desert soils	29	Aridisols, Entisols	Cambisols, Leptosols, Calcisols, Arenosols, Fluvisols
7.	Forest and Hill soils	-	Inceptisols, Alfisols, Mollisols, Ultisols, Entisols	Cambisols, Luvisols, Leptosols, Phaeozems
8.	Peaty and Marshy soil	-	Histosols, Inceptisols, Entisols	Gleysols, Cambisls, Fluvisols

The following are the major soil groups of India which are discussed with their salient properties and equivalents in the US comprehensive system of soil classification.

i) Alluvial soils

ii) Black soils

iii) Red soils

iv) Laterite and lateritic soils

In addition to the above some other important soil groups and

v) Saline and Alkali soils

vi) Desert soils

vii) Forest and Hill soils

viii) Peaty and Marshy soils.

i) Alluvial soils

Concept and Formation

The name *Alluvial* is given to the soils that have developed on *alluvium* irrespective of their place of occurrence and degree of profile development. The Alluvial soils are the least developed (A-C) to well-developed soils profile (A-B-C) The parent material of these soils (alluvium) is of recent origin and has been derived from the erosion products brought and laid down by the sea currents; the deltaic alluvium by rivers. These soils have weak profile development.

Distribution

They are extensively distributed in the states of Punjab, Haryana, Uttar Pradesh, Uttaranchal, Bihar, West Bengal, Assam and the coastal regions and occupy an estimated area of 75 Mha in the Indo-Gangetic Plaints and Brahmaputra Valley alone.

Salient Characteristics

- Alluvial soils are variable in colour and texture, depending on the source of parent material and their place of deposition. They are coarser near the source and become finer in proximity to the sea
- They are fluvial (stratified) in nature and that is reflected in their texture and irregular distribution of organic matter with depth.
- They are in general deep to very deep in nature

- Depending on the source of parent material and climate, these soils are generally calcareous and alkaline or acidic
- They exhibit different profile development from the least developed to well developed profiles
- They are inherently rich in plant nutrients. In general, they are fairly sufficient in phosphorus and potassium, but are deficient in nitrogen and organic matter content.
- The soils are prone to develop salinity and / or alkalinity wherever artificial irrigation is introduced, and / or the groundwater is brackish.

Classification

In the Genetic System, these soils are classified as Alluvial Soils, belonging to the Order Azonal. Alluvial Soil covers the following orders: Entisols (with A-C profiles), Inceptisols (with A-(B)-C profiles), Alfisols (with A-Bt-C profiles) and Aridisols (with A-(B)-C profiles).

Land Use

These soils are one of the best agricultural soils and are used for growing most crops. The major constraints of these soils are: *stratification* that restricts leaching and drainage and extreme sandy nature that results in excessive leaching of water and plant nutrients, *hydromorphic* condition that promotes reduction and results in poor aeration for plant growth. These soils, if managed well, can be fruitfully used for growing most of the agricultural and vegetable crops with high productivity. For sustainable production, these soils need artificial irrigation and judicious use of fertilizers and organic manures. These soils are suitable for wheat, rice, maize, groundnut and berseem, sugarcane, jute, sunflower, cotton etc.

ii) Black (Cotton) soils

Concept and Formation

The name black is given to soils that are very dark in colour and turn extremely hard on drying and sticky and plastic on wetting, and hence are very difficult to cultivate and manage. These are comparable with *Grumusols* of the USA. These are developed from basaltic parent material. Expansion and contraction set up a steady churning process in the pedon. Churning causes vertical mixing in deep soils and leads to the development of typical features, such as deep (> 50 cm) and wide (>1 cm) cracks, *gilgai* micro-relief and / closely intersecting *slicken–sides*.

Distribution

These soils are dominantly distributed in the intertropical and sub tropical parts of the world especially Africa, Australia, SW USA and India. These soils are predominant in the Central, Western and Southern states of India. Black (cotton) soils occupy an area of 74 Mha in India alone.

Salient Characteristics

- These are highly clayey with clay content ranging from 30 – 80 %
- Being calcareous, they have pH ranging from 7.8 to 8.7, which may go upto 9.5 under sodic conditions
- Being rich in smectitic clay minerals, they have high cation exchange capacity (30-60 C mol (P^+) kg^{-1} soil), have swell-shrink nature.
- They have high water and nutrient holding capacity. The moisture holding capacity, although is high (150-250 mm/m), yet a large part of it is not available for plant growth because it is held tenaciously by the dominant smectitic clay
- Because of high clay content and smectite, these soils pose drainage problem.
- These soils donot exhibit any eluviations and illuviation processes, because of churning.
- These have high bulk density (1.5 to 1.8 Mg m^{-3}) because of the swell-shrink nature of the soil
- Very dark in colour which may be due to clay-humus complexes and the presence of titaniferous magnetite material
- These soils are highly sticky and plastic and thus are difficult to manage and cultivate
- They pose many problems as sinking of floors, development of cracks in buildings, breaking and / or tilting of gas / water pipelines, tilting of electric and telecommunication poles, uprooting and / or tilting of trees, poles etc.

Classification

In the Genetic System, these soils are classified in the orders Intrazonal (Regur) and Azonal (Regosols, Alluvial soils). Black soils belong to the order Vertisols and their shallow members to the order Inceptisol, Entisol.

Land Use

The major constraints of these soils are low infiltration rate, poor drainage condition, poor moisture and nutrient supplying capacity for plant growth, poor in available plant nutrients, especially nitrogen, phosphorus and micronutrients. Due to swell-shrink nature, these are unsuitable for laying foundations, construction of buildings, laying of pipelines and electric communication poles etc. The soils are inherently very fertile and under rainfed conditions, they are cultivated for cotton, sorghum, millets, soybean, pigeonpea etc., Under irrigated conditions, they can be used for a variety of other crops, such as sugarcane, wheat and citrus plantation. These soils need careful management practices before planning their use for agricultural and non –agricultural purposes.

iii) Red Soils

Concept and Formation

The name Red is given to the soils that are rich in sesquioxides and have hue values of 7.5 YR or redder. These soils are developed on granite, gneiss on well drained, stable, higher landforms under hot, semi-arid to humid subtropical climatic condition. Weathering is moderately intense that leads to enhanced decalcification and mobilization of clay to form a clay enriched B horizon. Weathering products are leached out, leaving behind the less- mobile elements, like silica, iron and alumina. Iron and aluminium under oxidized conditions, form sesquioxides, imparting red colour to these soils.

Distribution

These soils occupy an area of about 70 Mha and are predominantly observed in the southern parts of the Peninsula, comprising the states of Andhra Pradesh, Tamil Nadu, Karnataka, Maharashtra, Orissa and Goa and the NE states of India.

Salient Characteristics

- These are red to yellowish in colour which is the result of coating of ferric oxides on soil ped surfaces. It is red when ferric oxides occurs as haematite or anhydrous FeO and yellow when it occurs as limonite
- The texture varies from loam to clay loam in texture
- They are shallow, gravelly and poor in soil fertility in the uplands to very deep, fertile in the plains and valleys
- They are well drained depending upon the topographic position and texture.
- They are neutral to acidic in reaction depending upon the content of iron oxides
- Silica / sesquioxide ratio varies from 2.5 to 3.0; the amounts of iron and aluminium are generally high (25-30%)

- The cation exchange capacity and base saturation in these soils are relatively lower than that in the black or alluvial soils. The CEC ranges from 35-50 cmol (p^+) kg^{-1} soil.
- These soils are generally deficient in nitrogen, phosphorus and potassium. They are also poor in organic matter and lime contents.
- They have compacted subsoil due to illuviated clay that improves the nutrient and water holding capacity, but poses physical restriction for root development
- These soils show the dominant presence of illite and chlorite with common occurrence of kaolinitie (1:1 type) clay minerals.

Classification

In the Genetic System, they qualify as red loam, reddish and yellowish brown soils. They belong to the orders: Alfisols (base-rich), Ultisols (well developed soils with low base saturation), Entisols (shallow and crusty) and Inceptisols (moderately developed).

Land Use

The soils pose limitations of soil depth (on hill and hill slopes), poor water content and nutrient holding capacity, surface crusting and hardening, excessive drainage and run off and poor natural fertility (in respect of N, P, Ca, Zn and S). Under good management practices, these soils can be used for a variety of agricultural, horticultural and plantation crops such as millets, rice, groundnut, maize, soybean, pigeonpea, greengram, jute, tea, cashew, cocoa, grapes, banana, papaya and mango.

iv) Laterite and Lateritic soils

Concept and Formation

The term laterite was used by Buchanan in 1807 for the highly ferruginous, vesicular and unstratified material observed in Malabar hills of South India. They are extremely weathered soils enriched in secondary forms of iron and aluminium and devoid of bases and primary minerals. The laterites are typically formed in tropical climate, experiencing alternate wet and dry seasons. Silica is leached during weathering under high temperature and the sesquioxides are left behind. On drying, these are converted into irreversible iron and aluminium oxides. These are hard or capable of hardening like bricks, when exposed to drying after wetting.

Lateritic soils are formed under almost comparable climatic conditions, but do not require alternate wet and dry conditions and the groundwater level may not be very near the surface. Lateritic soils are widely distributed in the states of Maharashtra, Karnataka, Tamil Nadu and NE states and occupy about 25 Mha of the total geographical area in the country.

Distribution

These soils are observed on hill-tops and plateau landforms of Orissa, Kerala, Tamil Nadu etc., and cover an area of about 40 Mha. Laterites are occasionally observed, whereas the lateritic soils are more extensively distributed.

Salient Characteristics

- The laterites have reddish colour (7.5 YR or redder) with maximum intensity in the B horizon, that decreases down the profile
- They are deeply weathered and the depth of weathering may extend upto a few metres
- They have high clay content (B horizon) which is no due to illuviation but due to insitu alteration of weatherable minerals
- They lose bases and silica with relative accumulation of sesquioxides
- The soils develop acidic reaction
- They are characterized by low silica : sesquioxides of less than 2
- The dominant clay mineral is kaolinite
- The major limitations include deficiency of phosphate due to high phosphate fixing capacity, high acidity, toxicity of aluminium and manganese and deficiency of potassium, calcium, magnesium , zinc and boron.
- The base saturation is < 40 % for laterites and alfisols in lateritic soils have high (> 35 %) and ultisols have low (< 35 %) base status.

Laterite Vs Lateritic Soil – A comparison

Laterites	Lateritic
Represent the order Oxisol	Represent the order Ultisol
BSP <40	BSP > 40
$SiO_2 / Al_2O_3 < 2$	$SiO_2 / Al_2O_3 > 2$
Clay mineral – 1:1 type	Mixed 1:1 and 2:1
Gibbsite present	Gibbsite absent

Classification

In the Genetic System, they qualify as yellowish brown or reddish brown soils. They belong to the orders: Oxisols and Ultisols.

Land Use

The major limitations include deficiency of P,K,Ca, Zn, B etc., high acidity and toxicity of aluminium and manganese. The low level laterites are used for growing rice, banana, coconut and arecanut and the high level laterites for cocoa, cashew, tea, coffee, rubber etc.,. The true laterites are occasionally used for cashew plantation crop. Shifting agriculture is practiced but the shifting cycle should be of 20 years or more inorder to restore the fertility status. Lateritic material, after dehydration is used as bricks for building purpose.

v) Desert (Arid) soils

Concept and Formation

The name desert / arid is given to the areas supporting almost negligible vegetation, except xerophytic plants unless irrigated. The soils developed in such dry regions representing *aridic* or *torric* moisture regimes are termed as Desert or Arid soils. They occur in all temperature regimes including Cryic, Frigid, hyper or megathermic. Because of the limiting rainfall (< 300 mm per annum), the soils show the accumulation of salts at or near the surface forming gypsic, calcic and salic horizon.

Distribution

A large tract of the hot and arid region, with a plant growing period of less than 60 days in a year and situated in the NW India between the Aravalli hills and the Indus river, have desertic conditions of recent origin. These soils occupy 29 m.ha in the total geographical area.

Salient Characteristics

- The soils are sandy to loamy fine sand in texture, with clay content ranging from about 3.5 to 10 percent
- These are pale brown to yellowish brown in colour and are either single grained or have weak , fine subangular blocky structure
- They are slightly to moderately alkaline in reaction (pH 7.8 -9.0) because of their calcareous nature. Calcitic nodules are present

- The hyper aridic areas such as Bikaner, Jaisalmer in Rajasthan are gypsiferous in nature and may form a gypsic horizon in the subsurface soils. They need special attention to avoid formation of sink holes, if irrigated for growing crops.
- These soils are poor in nutrient and water holding capacity because of their sandy nature.
- These are poor in soil fertility status, especially N,P,K,S and Zn, because of the sand cover, rich in quartz and poor biological activity.
- Weatherable minerals like Feldspars although present, donot weather because of the limiting rainfall.

Classification

In the Genetic System, they are classified as Pedocals, Sierozems, Calcareous Sierozems or Desert soils, belonging to the order: Zonal soils. Sand dunes of the arid region are classified as Regosols within the order Azonal. They belong to the orders: Aridisols and Entisols.

Land Use

These soils are fertile and are comparable with the Alluvial soils, except for aridity, that restricts their use for raising crops because of water constraint. Due to aeolian action, the dry sand move in the direction of the wind and poses a threat to the adjacent cultivated areas by depositing a thick mantle of sand at the surface. Sandy texture has poor water and nutrient holding capacity and excessive infiltration and leaching or drainage. Sink–holes is a problem in gypsic horizon.

These soils are inherently fertile because of weatherable minerals but because of aridity, the chemical weathering is limited to release cations. Once irrigated, these soils have the potential to grow two crops in a year. These soils can be profitably used for growing one to two crops in a year if irrigation is made available.

vi) Forest and Hill soils

This name is implied for soils developed under any forest cover. These soils are found in the places *viz.,* Himachal Pradesh, Jammu and Kashmir, Uttar Pradesh, Nagaland, Assam, Meghalaya, Madhya Pradesh,Mizoram and Manipur. These soils are two types *viz.,* podzolic and brown forest. (Area 75 m.ha).The major soils observed in different forest areas are Brown forest and podsolic (in Northern Himalayas), Black cotton, Red and lateritic (Deccan Plateau). Important soils in the northern and north-western Himalayas are.

a) Podzolic soils

Concept and Formation

These soils are formed under coniferous vegetation, in the presence of acid humus and low base status, show some characteristics associated with Podzolisation, but because of the unfavourable parent material, the process of podzolization is restricted only upto the mobilization of sesquioxides. Hence typical Podzols in northern Himalayas are not formed.

Salient Characteristics

- The soils are moderately to strongly acidic in reaction (pH 4.5 -6.0) and show moderate to well developed profiles with an argillic horizon (Bt) or a cambic horizon
- They are high in organic matter (2.0-3.5 %) and low base saturation (<50 %)
- These are variable in CEC (10-15 cmol (p^+) kg^{-1} soil). The clay content varies from 20-30 %
- These soils are deficient in phosphorus since it gets precipitated as iron and/or aluminium phosphates.

Classification

In the Genetic System, the soils developed under pinus vegetation and cool humid conditions qualify for Zonal (podzolic soils) and Azonal (Lithosols and Alluvial soils). They belong to the orders Inceptisols, Alfisols, Entisols and Ultisols based on topographic positions.

Land Use

These soils face a problem of erosion by water and/or deficiency of P_2O_5. They are used for growing a variety of crops like rice, maize, soybean, wheat, tea, fruit plants such as plum, pear, apple etc.,

b) Non – calcic brown forest soils

Concept and Formation

These soils develop from sedimentary rocks and / or alluvium or colluviums under sub humid to humid climate and mixed vegetation. They are rich in organic matter and high in base status.

Salient Characteristics

- The soils are deep in valleys and shallow on slopes
- They are neutral to slightly acidic in reaction (pH 6.0-7.0); the calcareous members have a pH around 8.2
- These are moderate to high in organic matter (2-3%) in the sub surface horizon which decreases with depth. They show high biological activity.
- They are moderate in CEC (15-20 c mol (p^+) kg^{-1} soil) and the exchange complex is saturated with bases (70-90 %)
- The soils have a mollic or mollic like ochric epipedon underlain by a cambic (or an argillic) B-horizon.

Classification

In the Genetic System, these soils are called Brown Forest and Non Calcic Brown soils of the order Zonal soils. They belong to the orders Inceptisols, Mollisols, Entisols and Alfisols.

Land Use

These soils are very productive and have great potential for growing agricultural and plantation crops such as rice, maize, soybean and fruit plants such as apple, almond, pear, apricot etc. The steep sloping lands are kept under natural vegetation.

vii) Salt affected soils

Concept and Formation

The term **'salt affected'** is given to the soils that show accumulation of soluble salt and have sodium on the exchange complex that impairs the growth of most plant species. Sodic soils occupy lower topographic positions, where the products of weathering accumulate during the monsoon rains by surface runoff. In the post rainy period, the water evaporates and the soil solution becomes concentrated resulting in increased SAR, ESP and pH. Saline soils of the coastal regions develop from the brackish ground water due to capillary action under excessive evaporation demand.

Distribution

These soils occupy about 10 M ha of which 7 Mha is sodic and the rest is saline (CSSRI, Karnal).

Salient Characteristics

Saline soils

- They have high soluble salts of chlorides and sulphates of sodium, calcium and magnesium. Electrical conductivity is >4 dSm^{-1}
- Exchangeable sodium percentage is < 15
- pH < 8.5 (due to calcium carbonate)

Sodic soils

- They have an electrical conductivity is < 4 dSm^{-1}
- Exchangeable sodium percentage is > 15
- pH > 8.5 (9.0 -10.2)
- Greyish colour because of poor drainage
- Platy surface and blocky sub-surface soil structures.
- Low infiltration and permeability rates because of the dense B-horizon resulting from the flocculation and migration of clay filling the pore spaces.

Classification

Salt affected soils were observed since ancient times (2;500 BC) when such soils were termed as Usar. In the Genetic System, they were classified as Solonchaks (saline) and /or Solonetz (sodic). They belong to the orders Aridisols, Inceptisols and Alfisols.

Land Use

The sodic soil poses serious problems of high sodium on the exchange complex and poor physical conditions, especially soil structure and drainage, nutrient and water availability and micronutrient deficiency. Sodic soils once ameliorated using gypsum technology is suitable for growing a variety of crops.

viii) Peat and Marsh soils

Concept and Formation

The name Peat is given to soils having large accumulation of organic matter developed under humic climate in low –lying, coastal marshy land or soils occupying lower topographic positions. Such soils occupy a limited area in localized pockets of Kerala, Tamil Nadu and NE states.

The Marshy soils is the name given to the soils that are confined to depressions caused by dried lakes in coastal plain areas. The soils develop under anaerobic, waterlogged environment and result in the formation of bluish / greenish soils because of the reducing conditions. Such soils containing some soluble salts are termed as Kari in Kerala. The soils remain submerged in water during the monsoon rainy season. In SE Asia, such soils are termed as Cat Clay or Acid Sulphate soils.

Distribution

These soils are occasionally observe in the coastal intertropical lowlands. Such soils occupy a limited area of about 12Mha, largely in SE Asia, West Africa and S.America. In India they are found in Kerala, Orissa, Sunderbans in West Bengal, SE Tamil Nadu and NE states, especially Tripura, where they support mangrove vegetation.

Salient Characteristics

- These soils are dark to almost black in colour with abundant organic matter (20 - 40 %)
- Very fine in texture
- Very strongly acidic in reaction (pH 3.5 - 4.0) due to decomposition of organic matter under anaerobic conditions where no nitrification is feasible.
- Show accumulation of ferrous and aluminium sulphates, iron pyrites especially in the tidal swamp areas. When drained, the pyrite (FeS_2) is oxidized for form sulphuric acid, resulting in significant drop in pH values.
- Have clay minerals *viz.,* kaolinite and smectite.

Classification

Such soils are termed as Cat- Clay or acid sulphate soils in SE Asia, West Africa, S.America and N.Europe especially Scandinavian countries. In India, such saline peat soils are termed as Kari in Kerala. They belong to the orders Inceptisols and Entisols.

Land Use

These soils are problematic because of the presence of pyrite, toxicity of iron and aluminium, deficiency of P, subsidence of organic matter on drainage making drainage operation difficult. Because of these constraints, these soils are unsuitable for the cultivation of most crops. Such soils in Kerala are cultivated for rice by draining the flood water received during the monsoon rainy season.

In Sunderban areas of West Bengal, such soils are maintained under mangrove forest that can sustain strong acidity. In Vietnam, pine apple is grown on the top of the slopes of the ridges.

3.8. Soils of Tamil Nadu

Tamil Nadu is located in the south of India and covers an area of about 3 million ha. The state is divided into four geomorphologic zones, *viz.,* the coastal plain, the Eastern Ghats, the central plateau and the Western Ghats. The coastal plain stretches from Pulicat lake in the north to Cape Camorin (kanyakumari) in the south. The coastal plain is backed by a broken line of hills *viz.,* The Javadus, Shervaroys, Kalrayans, Pachamalai and Kollimalais. This line of hills is known as the Eastern Ghats. On the western border occurs a group of high hills known as Western Ghats. They are the Nilgris, Anamalais, Palani and cardamom hills. Between the Eastern and Western Ghats lies the central Plateau with elevations ranging from 152 to 610 metres above mean sea level. The general topography is undulating with an overall sloping from west to east.

Classification	:	Semi-arid (tropical monsoon)
Temperature regime	:	Coast – Isohyperthermic
		Inland – Isohyperthermic
		Hills – Isothermic
Moisture regime	:	Coast and Inland – Ustic
		Hills- Udic

The major soil groups of Tamil Nadu

Red soils (62 per cent)	The red soils are further classified
Alluvial soils (16 per cent)	as red loamy (30 per cent), redsterile (6 per cent),
Black soils (12 per cent)	red sandy (6 per cent),
Laterite soils (3 per cent)	thin red (12 per cent)
Coastal soils(7 per cent)	& deep red loamy soils (8 per cent).

As per the USDA system of soil classification (Soil Taxonomy), the soils of Tamil Nadu are classified into six orders, *viz*., Entisols, Inceptisols, Alfisols, Mollisols, Ultisols and Vertisols.There are 12 suborders, 20 great groups, 44 subgroups, and 94 soil families.

About 50 per cent of the total area of state is occupied by Inceptisols, 30 per cent by Alfisols, 7 per cent of Vertisols, 6 per cent by Entisols, 1 per cent by Ultisols and a very negligible area under Mollisols.

i) Entisols (6.1 %)

- These soils cover an area of 18,28,978 ha and are distributed in Nilgiris, Erode, Pudukottai, Ramnad, Sivagangai, Salem, Trichy, Tirunelveli, Thanjavur, Coimbatore, Dharmapuri and Kanyakumari districts.
- River alluvium, coastal alluvium and eroded soils are included. These are characterized by sandy texture or fine texture alternated with sandy layers.
- They are poor in nitrogen, phosphorus and organic matter but rich in potassium and lime.
- The CEC is 25 C mol (p^+) kg^{-1} soil and the silica sesquioxide ratio is 12.5.
- The dominant clay mineral is 2:1 type (illite).
- River alluvium is used for the cultivation of wetland crops like paddy, sugarcane and banana; Coastal alluvium is used for casuarina plantations. Eroded areas are used for the development of pastures.

ii) Inceptisols (50.3%)

- These soils are distributed in all the districts of Tamil Nadu and cover an area of 22,10,685 ha
- Moderately deep red, brown and black soils are included under this category.
- The soils are poor in nitrogen, phosphorus, potassium and lime.
- They have moderately well developed subsoil. The soils are rich in kaolinitic type of minerals
- The CEC is 10 to 15 C mol (p^+) kg^{-1} soil
- The soils are used for cultivating sorghum, groundnut, maize, onion, tobacco and chillies.

iii) Alfisols (30.3%)

- These soils cover an area of 31,43,312 ha and are distributed in all the districts except in the hill ranges.
- The soils are very deep, reddish in colour and have a well developed sub surface horizon
- They have a free drainage and the pH ranges from 6.5 to 8.0.
- The soils are generally low in total soluble salts.
- Cation exchange capacity is 10-15 C mol (p^+) kg^{-1} soil. The soils contain low N and P_2O_5 and medium to high K_2O

- The soils are utilized for cultivation of millets and pulses under dryland conditions and groundnut, cotton, maize, sugarcane and paddy under irrigated conditions.

iv) Ultisols (1%)

- These soils cover an area of 36,499 ha and are distributed in Salem, Dharmapuri and Nilgiris districts
- The soils are very deep and highly weathered. The soils are dark coloured in the surfaces due to high organic matter content (2-5 %).
- They are acidic in soil reaction. The CEC is 8-15 C mol (p^+) kg^{-1} soil.
- The soils are characterized by low amounts of silica and high amount of sesquioxides. Silica –sesquioxide ratio is below 1.33.
- The soils are poor in bases. Phosphorus is largely unavailable due to fixation in the soils are iron and aluminium phosphates.
- These soils are used for the cultivation of coffee, tea, cocoa, cold vegetables *etc.,*

v) Vertisols (7.0%)

- These soils cover an area of 17,91,575 ha and are distributed in all districts except in Nilgiris and Kanyakumari districts.
- Deep black cotton soils and old alluvial soils are included in this order.
- These soils are very deep, clayey, calcareous and poorly drained. They develop cracks during summer.
- They are poor in organic matter, nitrogen and phosphorus but fairly well supplied in potassium and lime.
- They contain high amounts of iron, calcium and magnesium.
- The soil reaction is mildly alkaline (pH 7.4 -8.2). The CEC is also high *viz.,* 30-70 C mol (p^+) kg^{-1} soil.
- Free calcium carbonate is present in the form of canker.
- Soils are cultivated with cotton, sorghum and pulses under rainfed conditions. Under irrigated conditions, paddy and cotton are grown.

vi) Mollisols (0.3%)

- A negligible area is covered under Mollisols. It is rich in organic matter.

3.9. Soil Maps

Soil map is a map in which mapping units are based specifically on soil profile. A map based on classes such as "soil-land form association" is a soil map if it is directed towards showing the distribution of soils. A geographical representation showing diversity of soil types or soil properties. The geomorphological map is primarily a map of land form units with soils being added to the legend.

3.9.1. Kinds of Soil Maps and their Preparation

1. Point soil maps

Maps where the actual sample points are shown along with their soil class or one or more properties. This map gives direct representations of what was actually sampled. But all the area is not covered.No spatial variability is implied.

2. Area – class polygon maps

Survey area is divided into polygons by precise boundary lines, each polygon being labeled with a class name and each class in turn being described by a legend. Soil survey maps of this type can easily represented by Vector GIS model.

3. Continuous field maps by interpolation

These maps are made by interpolation using raster GIS model. These maps show the continuous distribution of soil properties. There are no sharp boundaries and all the variations are said to be continuous.

4. Continuous field maps by direct observation

The measurement is made at every point and commonly presented in a grid map using raster GIS model. These maps show measured continuous distribution of soil properties. Currently there is much interest for this map in precision farming. Common examples are height and vegetation index. Sensors are satellites, air craft or ground traverse.

Significance of Soil Maps

- For practical purposes the areas of relative's uniformity of soils are shown on the maps. They are delineated by lines which are placed where the rate of change is more rapid.
- The soil maps form the basis to develop maps for a wide variety of uses. The area can be grouped into land capability classes.

- Soil conservation service personnel use these maps to develop conservation plans for farmers.
- Banks, insurance companies and other money fending agencies uses soil maps for determining security for loans.
- Real estate companies and individuals interested by buying or selling land make extensive use of soil maps.
- These are also used by highway, irrigation and drainage engineers.

Types of soil maps

1. General maps: (Scale less than 1:1000000)

These maps do not claim to be very accurate; they reflect the general features of soil cover the territories of individual states (or) district natural regions. Basically they are compiled for our students. The mapping units delineated on these maps are **associations of series**. These generalized maps usually are made by combining the delineations of detailed soil survey maps to form broader map units.

2. Small scale maps (Scale is 1:1000000 to 1:300000)

These maps are compiled for departments (provinces) or regions. The aim of this map is to assess the land resources, to plan the rational ways of their proper use and development, to select soil areas for priority land management.

3. Medium scale maps: (Scale is 1:300000: 1:100000)

These types of maps are compiled chiefly for administrative units (regions, district etc).They are designed for planning the locations of agricultural enterprises such as cooperative forms, state and the area requiring amelioration.

4. Large scale maps: (Scale from 1: 50,000 to 1:10000)

These are mainly compiled for smallest administrative units such as taluks or blocks. They are used as guidance for improving soil fertility for more rational and effective use of soils etc.

5. Detailed maps: (Scale from 1:10,000 to 1: 2000)

These maps are compiled for the territory of experimentation stations, plantations of perennial crops and for agricultural objects which must accurately account for the differences in soils.

Types of maps based on utility

Generalized Soil Maps

Generalized soil maps are made to reveal geographic relationships that cannot be seen readily on detailed maps. Most soil survey reports include a general soil map for the area. The scale of these maps depends on the intended uses. A detailed map is generalized by enclosing the larger areas within which a few kinds of soil predominate in relatively consistent proportions and patterns. These larger areas are described in terms of the dominant soils. Areas defined as associations of **soil series or their phases are combined in this process into larger areas** that can be defined in terms of associations of taxa at higher categories. They merge with the 4th order soil survey – Detailed Soil Survey.

Generalised soil maps of states

Many state general soil maps have been published at scales of 1:1,000,000 to 1:15,00,000. Map units generally are associations of soil series, although associations of higher taxa have been used on some maps. These general soil maps provide an overview of the distribution of the more extensive soils of the state. They are useful in broad land use planning.

Regional and National Soil Maps

The scale is commonly smaller than 1:1,000,000. Units on regional and national maps are usually associations of **great groups or suborders**. Some regional maps have accompanying booklets. National soil maps are used for studying very broadly defined capabilities and limitations that affect regional and national issues.

Uses of generalized maps

The possible uses for generalized soil maps are for appraising the basic soil resources of whole countries, for assisting farm and for guiding commercial interests. These maps are compiled for country and regional land-use planning. They are useful in predicting the general suitability of large areas of soils for residential, recreational, wildlife, and other nonfarm uses, as well as for agriculture. It helps to suggest alternative routes for roads and pipelines where the least problems with soils are expected.

Schematic Soil Maps

Schematic soil maps differ from generalized soil maps as it is compiled from information other than pre-existing soil maps. Scale is commonly 1:1,000,000 or smaller, although useful maps are sometimes made at larger scales. Schematic soil maps are commonly made as a preliminary step to locate areas where further investigation is justified. For undeveloped regions, a schematic soil map is useful in advance of an organized field survey. Information about climate, vegetation, geology, landforms, and other factors related to soil are gathered and studied. Data obtained by remote sensing techniques, including aerial photography, may provide useful information. Schematic soil maps merge with 5th-order (exploratory) soil surveys.

Preparation of soil maps

The soil survey methods have undergone large changes during the recent past with the availability of satellite imagery, apart from topographical maps and aerial photographs.

Soil survey and mapping involves the following steps

- Preliminary reconnaissance of the area to investigate the major soils and their pattern of occurrence.
- Procurement of required base maps. Aerial photographs, satellite imagery, topographical and other maps are useful references and are used as the mapping base.
- Preparation of mapping legend based on the preliminary field studies.The mapping legends are modified as the survey progress.
- Stereoscopic study of aeriol photographs for the identification and delineation of land forms (hills, valley, terraces, flood plains, coastal plain and sand dunes.) and their sub-divisions based on the differences in tone, relief and vegetation
- The differences in tone or colour serve as the basis for drawing tentative boundaries and for predicting kinds of soils. The predictions are verified in the field, but preliminary interpretation can increase the quality of mapping.
- The soils are examined at some standard interval along the traverse to locate important difference in soil properties.
- Interval between traverses and soil observation points cannot be specified. It depends on the scale of mapping and variability of soil properties.
- Plotting of soil boundaries, some soil boundaries are sharply defined, others are plotted as lines mid way in zones of gradual transition from one soil to another.
- The predicted boundaries are checked throughout their course in high intensity surveys and in other cases, the boundaries are plotted mostly by interpretation of remotely-sensed data and verified with some observations.

- Classification of soils and naming of map units.
- Preparation of final legend and finalization of soil map.

3.9.2 Soil Survey Report

Soil and land use surveys help in the assessment, mapping and proper interpretation of the basic data on soil and characteristics for various land use purposes. Soil survey can be regarded as complete only if report that describes the kinds of soil shown on the map and their capabilities for use is given to enable farmers, farm advisors, planners and other users to make full utility of the data.

Soil survey reports or soil bulletins are the end products of any survey. They form the essence of all field investigations.

Pre Requisites of Soil Survey Report

- For writing a soil survey report, different approaches may be followed depending upon the inclinations of the organization and the report writer.
- The survey report should be attractive and technically sound and convenient for the readers, most of whom are not soil specialists.
- Besides written text, the report should be illustrated by photographs, tables, diagrams and block diagrams.
- The report should meet the requirements of different end users viz., policy makers (development process), soil scientists (profile characteristics for improving agricultural productivity) and for others to know about the soils and their formation by themselves.

Outline of Report

All India Soil and Land Use Organization suggest the following outline for report writing.

1. Abstract
2. How to use soil survey report
3. Table of contents
4. Introduction
5. General description of the area
    a) Location and extent
    b) Physiography, relief and drainage
    c) Geology

e) Climate

f) Natural vegetation

g) Water supply

h) Socio-economic condition

6. Land use and agriculture

a) Present land use

b) Agronomic practices and crops

c) Forest

7. Soil survey methodology

8. Soils of the area

9. Soil survey interpretations

i) Thematic interpretative classification

ii) Land Capability Classification

iii) Land Irrigability Classification

iv) Land Suitability Classification

v) Productivity Classification

10. Problems and suggestions

Abstract

The abstract may include broad facts, conclusions and recommendations offered by the surveyors.

How to use soil survey report

This explains the reader how to locate the fields of his interest on the map viz., map scale and symbols for land marks.

Table of contents

The purpose of the table of contents is to help the reader to find specific items in the report.

Chapter 1. Introduction

Explains the purpose of the survey and the circumstances leading to it. It gives a brief account of the survey area, intensity of survey and the organization which conducted the survey. A simple location map giving all the places mentioned in the report should be provided at the beginning of the chapter.

Chapter 2. General description of the area

a) **Location and extent** -Latitude and longitude, boundaries, main geographic names like rivers, towns, main roads etc., are to be given.

b) **Climate** -Temperature and rainfall (precipitation) data in tabular form should be given. Information like frequency and severity of drought, detrimental wet periods and floods, form and intensity of rain, snow, hail, local variations of climate due to elevation, nearness to lakes and ombrothermic diagram should also be given.

c) **Physiography, relief and drainage** - If the area has two or more distinct physiographic divisions, these may be described giving the principal soils in each division.

d) **Geology** - Main geological formation of the area should be explained. Information on weathering of the parent material is essential for soil classification.

e) **Natural vegetation** - Trees, bushes and shrubs present on different soils in different land systems of the area.

f) **Water supply**- Sources of water for domestic use, livestock and irrigation.

g) Present land use and agriculture

- **Land use:** A general description on the kinds and areas of cropped fields, forests, orchards, grasslands etc., in the various system and land types
- **Agriculture:** Acreage and production of principal crops, crop rotations, area under different crops with the reasons, cultural practices, harvesting and storage may be explained.

h) Socio economic condition Settlement and population : Settlement, agricultural and industrial developments and its population

i) **Industries :** Kind and locations, effect on agriculture and rural population

ii) **Transportation and markets :** Available rail, bus, ship lines, aircraft and other transportation facilities in relation to agriculture particularly in relation to geography of soil.

iii) **Community facilities:** Schools, hospitals, libraries, telephones, electrification, recreation and other features related to rural development.

Chapter 3. Soil Survey Methodology

Details about the procedure followed for soil survey, base maps used etc.,

Chapter 4. Soils of the area

i) Series identified – All the soil series recognized in the area may be introduced.

ii) Soil genesis – Origin and character of the various pedogenic processes i.e., the characteristics of various soil horizons which have formed under the influence of present or paleo climatic conditions and living matter in combination with the relief, parent material etc.,

iii) General description of the soil series – Description of various soil series mapped is given. Each soil series is described highlighting the differentiating characters and other physical and chemical characteristics.

Chapter 5. Soil Survey Interpretation

The interpretation of the soil data is most important to the actual user, agricultural planner, the engineer and so on. The following interpretations are usually made from soil survey data.

a) Thematic interpretative classification

b) Land Capability classification

c) Land Irrigability classification

d) Land Suitability classification

e) Productivity classification

Chapter 6. Problems and Suggestions

Precise conclusions on nature of the soils, their distribution and their potential uses have to be given along with broad recommendations to guide the user.

Chapter 7. Annexures

Annexures are vital because they enable the surveyor to present essential data without mixing the main text. The following annexures are normally included.

i) Morphological description of the soil series

ii) Index to village wise mapping units

iii) Legend for soil symbols

iv) Analytical data of representative soil profiles

v) Glossary

Chapter 8. Maps

The soil and interpretative maps complete the soil survey reports. Maps must be clear, attractive and self explanatory. In detailed soil survey, mapping is done at phase level. Name of the soil series, effective depth, surface texture, slope, erosion intensity and surface characteristics such as stoniness, rockiness or salts are indicated. The map should have a legend which can easily be related to soil legend. Suitable colour coding should be followed in the maps. Light colours for good properties and dark colours for negative or poor properties. Indiscriminate use of colour and shading techniques has a negative effect.

Chapter 9. Summary and Conclusions

Individual soil related constraints should be enumerated first and accordingly suggestions for better management should be given. The report should end with a brief summary about the area of survey, type of survey, soils met along with their extent and the objectives of the survey.

The soil survey reports or the soils bulletins have been found to be important, particularly in designing drainage and irrigation networks, planning the ameliorative measures for salt affected and eroded areas, for sanctioning loans, for developing perspective land use plans etc. With an ever advancing technology, a soil map and report thought useful for all needs may require re-interpretation after 20 to 25 years depending upon the changing land use and/ or future needs. The soil resource maps also need resurvey for monitoring the soil health.

4

Soil Survey Interpretations

In recent times, the interpretations cover a wide spectrum of soil uses, both for agricultural and non agricultural (sewage and human waste disposal, high way construction, water and gas supply, recreation, wild life, urban and industrial structures) and many specialized uses. Soil survey if correctly carried out according to scientific principles, can provide the data base for various interpretations. This is because the same soil properties determine the behaviour of a soil for a variety of different uses. For instance, the content of clay, silt and sand, the shrink and swell behaviour, the permeability of soil to water and air, content of calcium carbonate, salts, organic matter, gypsum, soil depth, depth of ground water, hard pan formation, co-efficient of linear extensibility, water holding capacity *etc.* are examples of such properties that are important to plant growth and these properties also strongly influence non farm uses. Some examples in support are

A hard or petrocalcic horizon or other slowly permeable layer (clay-pan) in a soil restricts the movement of air and water, thus adversely affecting the plant root growth. Such a layer limits the soils suitability for septic tanks and other onsite waste disposal. It also makes a poor sub grade for high way construction.

A soil high in soluble salts is a poor medium for plant growth; it is also corrosive to iron and steel structures, such as pipelines and fences, causing high maintenance and repair costs.

High organic matter content is generally favourable to plant growth, but it is a limitation for road construction.

The strongly gypsiferous soils (gypsum content >10%) are considered unsuitable for arable crop cultivation because of its dissolution under irrigated or even moderate rainfall and making the surfaces uneven through the developments of sink holes. Such soils are unsuitable for irrigation and drainage canal networks and also for construction of roads, bridges etc.

The shrink- swell soils of India (vertisols) because of summer monsoon rains are suitable for growing a "kharif" crop but are very dangerous for construction of roads, air strips, buildings, laying gas, water pipelines which get distorted, develop cracks and lead to sunken floors etc.

Citrus orchards should not be planned in a soil with calcic sub surface horizon because of the possibility of lime induced iron chlorosis.

The application of soil survey depends on the quality of soil maps which is often evaluated by the purity of mapping units and the accuracy of map boundaries. To attain a high degree of accuracy, detailed mapping is required which can be very complex, expensive and time consuming. The factors influencing decision making on suitability of a soil for different uses are to be studied in a soil survey report and interpretations made accordingly for better land use.

4.1. Interpretative Groupings of Soils

4.2. Land Capability Classification

Land capability classification is the grouping of land units into defined classes based on their capability. Evaluation of land for land use planning is a consequent step following soil survey and mapping process. The capability of land units depends on their limitations and it is designed to emphasize on the hazards in different kinds of soils. It serves as a guide to assess the suitability of different land units for arable crops, grazing / forestry. This system however, may not be suitable for intensively cultivated plain areas, such as Indo Gangetic plain.

Depending on the nutritional and other properties soils behave differently. Some are suitable for cultivated crops, others for pastures and still others may be suitable only for forestry (or) recreation.

The grouping of soil units into capability classes and sub classes is done on the basis of their capability to produce crops and pasture biomass without adversely affecting the productivity over a long period of time. The criteria used for assessing a land unit are the physical land properties and the degree of limitation with which crop growth is inhibited. It is mainly based on

i) Inherent soil properties

ii) External land features

1. Environmental factors that limit land use (climate vegetative)

The first two information can be derived from a standard soil survey report and the third information is collected from other agencies (or) sources.

Factors determining the capability of a soil

- Depth of soils, stoniness, rockiness
- Texture and structure of soil
- Permeability (Movement of air and water through the soil)
- Relief (as expressed by slope)
- Extent of erosion
- Susceptibility to over flow, flooding and degree of wetness
- Presence of toxic salts, alkali and other unfavourable chemical properties such as pH, gypsum salts and aluminium.
- Climatic variation (in respect of soil temper5ature and moisture regimes)

The capability classification does not suggest the most profitable use of land. Further, these grouping are subject to changes as new information about the soils and their response become available.

The capability classification consists of three categories namely

1. Capability classes
2. Capability sub-classes
3. Capability units

1. Capability classes

Eight capability classes are recognized. The soils having greatest capability for response to management and least limitations are grouped in class – I and those having least, capability and greatest limitations are grouped in class-VIII. Each capability class is correlated with the cropping system as conditioned by LGP.

Land capability classifications (Eight classes)

Land suitable for cultivation				**Land not suitable for cultivation**			
Land capability class				**Land capability class**			
Class I	Class II	Class III	Class IV	Class V	Class VI	Class VII	Class VIII
Very good land (No limitation)	Good land (Minor limitation)	Moderately good (major limitation)	Fairly good (major limitation with occasional cultivation)	(No limi-tation) Suitable for pasture and grazing	(Minor limitation)	(Minor Limitation)	suitable for wild life and recreation

Brief description / salient features of LCC

Name of the class	Colour shown on land capability	Salient features	Crop suitability crops	Management practices
Class – I	Green	Excellent cultivable deep, nearly level. Productive land with no limitation (or very slight hazard)	Suitable for most crops like wheat, barley, cotton, maize tomato and beat	Need no management practices for cultivation
Class – II	Yellow	Good cultivable land on almost level plain or on gentle slopes limitation of soil depth, salinity, texture, drainage / erosion that reduces the choice of plants	Suitable for wheat, barley, cotton Moderately suitable for maize, alfalfa, tomato Slightly suitable for beans	Cultivate with precaution need simple management practices.
Class III	Brown	Moderately good cultivable land on almost level plain or on moderate slope. Limitation :L Moderate erosion soil depth, soil salinity and soil texture. They may have vertic characteristics (or) drainage problem that reduces the choice of crops	Unsuitable for growing vegetable crops. Varying suitability for different crops	Cultivation with careful management practices need intensive care.
Class IV	Pink	Fairly good land on almost level plains or moderately steep slopes **Limitations** Strong / very strong soil salinity S3/S4 shallow soil depth, strong erosion, fine soil texture. Poor (or) excessive drainage.	Limited cultivation unsuitable for growing a variety of crops. Suitable for selected crops 8 for pasture. Not economical to cultivate	Need intensive soil conservation and management practices
Class V	Dark grey	**Limitations** Story, rocky nature (or) marshy-ness use of implements is difficult	Not suitable for arable farming suitable for grazing	
Class VI	Orange	Moderate limitation such as steep slope severe erosion, limiting soil depth, strongly gypsiferous, stony or sandy (sand – done areas) eg. dense forest lands of the Himalays. Sandy (gypsiferous) and donal areas of Thar Desert in Rajasthan and SW Irag.	Non-arable land well suited for grazing (or) forestry	
Class VII	Red	Severe limitation such as steeply sloping land subjected to erosion (or) very shallow stony soils that have not	Fairly well suited for grazing or forestry but not cultivable	Need careful management for grazing and forestry
Class VIII	Purple	Severe limitation Very steeply sloping, highly eroded, rocky barren mountain landscape, hyper arid rocky undulating surface	Non-arable, suitable for wildlife (or) recreation. Extremely rough, rocky, arid, wet (or) extremely saline land	

2. Capability sub classes

The capability subclasses are based on kinds of dominant limitations such as wetness or excess water (w), Climate (c), soil (s) and erosion (e). The subclasses are mapped by adding limitation symbols to the capability class number subscripts for example II e, IIIw etc. Therefore the subclasses indicate both the degree and kind of limitations. The capability subclasses provide information as to the kind of conservation problems or limitations involved. There are no sub-classes in capability class-I land, since there is no limitation in this class.

3. Capability units

A capability unit includes soils which are sufficiently uniform in their characteristics, potential and limitations and require fairly uniform conservation treatment and management practices. If the existing limitations can be permanently removed (or) reduced in extent by proper conservation measures (or) management practices, such as provision of irrigation, drainage, control of gullies and construction of flood – retarding irrigation structures, then the land capability class can be changed towards a better class. A further deterioration in the existing conditions will shift the capability to a worse class.

Land Capability Classification- Quantification of the Criteria

Characteristics	Class I	Class II	Class III	Class IV	Class V	Class VI	Class VII	Class VIII
Topography (t)								
Slope (%)	0-2	3-6	7-12	13-25	0-2	25-35	35-50	50-60
Wetness (w)	No (F0)	No(F0)	No(F0)	No to slight (F1)	No to severe (F3)	No to severe (F3)	No to severe (F4)	No to severe (F4)
Flooding (%)								
Drainage	Well drained	Moderately Well drained	Imperfectly drained	Poorly drained	Very Poorly drained	Excessively drained	Excessively drained	Excessively drained
Physical soil conditions								
Surface texture	Loam	Sil & cl	Sl & C	Scl	S	ls	ls, s	ls, s
Surface coarse fragments (vol %)	1-3	3-15	15-35	35-50	35-55	50-55	56-75	75-85
Surface stoniness (%)	None	<0.01	<0.1	<0.3	<15	<15	<75	<75
Subsurface coarse fragments	0-15	15-35	35-55	55-75	55-75	55-75	>75	>75
Soil depth	100-150	50-100	25-50	15-25	15-25	10-15	10-15	0-10
Fertility	16-40	12-16	12-16	-	-	-	-	-
Apparent CEC								
Base saturation	80-98	50-80	35-50	15-35	15-35	15-35	10-15	0-10
O.C (0-15 cm)	1.0-1.5	0.75-1.0	0.60-0.75	0.4-0.6	0.4-0.6	0.4	0.3-0.4	0.3-0.4

4.3. Fertility Capability Classification

The fertility capability classification (FCC) was developed by Buol, Sanchez and co-workers (Buol,1972; Buol et al, 1975, Sanchez et al, 1982) as a technical system for grouping soils according to the kind of problems they present for agronomic management of their chemical and physical properties. The system emphasizes quantifiable topsoil parameters as well as subsoil properties directly relevant to plant growth and yield performance.

The FCC system consists of three categorical levels

Type (topsoil texture),Substrata type (subsoil texture) and Condition modifiers, including several changes from the original version (Buol et al, 1975). The classes within each categorical level are defined below. Class definitions from the three categorical levels are combined to form an FCC unit.

Type

The Type is the highest category, it is determined by the average texture of the ploughed layer or surface 20 cm, whichever is shallower.

S = Sandy topsoils- Loamy sands and sands (USDA)

A = Coarse loamy top soils- Sandyloams

M = Fine loamy top soils – Sandy clay loams and loams

T = Silty loamy topsoils silt, silt loam, silty clay loam and silty clays

C = Fine clayey topsoils > 35 % clay – sandy clay to clay

H = Very fine clayey top soils - > 60 % clay- heavy clay

O = Organic soils > 30 % O.M to a depth of 30 cm or more

Substrata Type

The Substrata Type is the texture of the subsoil that occurs within 50 cm of the surface. It is used if the subsoil texture differs from that of the surface (Type) within the defined limits. If no textural change of this magnitude is present, no Substrata type designation is employed.

S = Sandy topsoils- texture as in type (Loamy sands and sands)

A = Coarse loamy top soils- Sandyloams

M = Fine loamy top soils – Sandy clay loams and loams

T = Silty loamy topsoils silt, silt loam, silty clay loam and silty clays

C = Fine clayey topsoils > 35 % clay – sandy clay to clay

H = Very fine clayey top soils - > 60 % clay- heavy clay

R = Rock or other root restricting layer

Condition Modifiers

The modifiers indicate specific fertility limitations with different possible interpretations. Where more than one criterion is listed for each modifier, only one needs to be met. The criterion listed first is the most desirable one and should be used if data are available. Condition modifiers are used as lower case letters for coding the soils.

g : This modifier refers to a gley condition in the soil as an indication of the presence of a water saturation within 60 cm of the surface during some part of the year.

d : This modifier refers to an annual dry season of at least 60 consecutive days. It is defined to roughly correspond to Ustic, Xeric, Torric, and Aridic moisture regimes of the U.S. Soil Taxoomy.

e : This modifier refers to soils with very low cation exchange capacities in the ploughed layer.

a : This modifier refers to high concentrations of aluminum which may be toxic to most agronomic crops. It also implies a high degree of phosphorus fixation by aluminum compounds.

h : This modifier refers to a moderate level of acidity that would retard the growth of some aluminum sensitive plants. Since "a" and "h" conditions are often altered by liming or by the residual acidity of various fertilizer sources, the soil should be examined to a depth of 50 cm.

i : This modifier intended for those soils where phosphorus fixation by iron compounds is of major importance.

x : This modifier attempts to delimit soils with allophane dominated mineralogy.

v : This modifier indicates clayey soils dominated by 2:1 expanding clays. The fertility implications are for high permanent charge CEC and difficulties in water management and soil tillage. It is thought that this modifier will be loosely aligned with the Vertisol order and some vertic subgroups.

k : Many soils have small quantities of potassium bearing minerals and profitable responses to potassium fertilizers are expected. This modifier attempts to delimit those soils where it is almost certain that potassium will be needed in an agronomic fertility program.

b : This modifier delimits calcareous soils or, more specifically, free carbonate within 50 cm and phosphorus fixation by calcium compounds.

s : This modifier separates those soils with sufficient salinity to present problems for most crops.

n : Sodium is considered because of its effect on clay dispersion and on moisture availability. This modifier is designed to delineate soils with a sodium problem.

c : This modifier indicates the presence of acid sulfate soils and the associated management problems.

---

g Gley; soil or mottles < 2 chroma within 60 cm of surface and below all A horizons or saturated with H_2O for > 60 days in most years

d Dry Ustic or xeric environment; dry > 60 consecutive days per year within 20-60 cm depth.

e Low CEC< 4 meq/100 soil by " bases + unbuffered Al< 7 meq/100 soil by " cations at pH 7< 10 meq/100 soil by " cations + Al + H at pH 8.2

a Aluminium toxicity > 60 % Al saturation of CEC by (" bases and unbuffered Al) within 50 cm> 67 % Al saturation of CEC by ("cations at pH 7) within 50 cm> 86 % Al saturation of CEC by ("cations at pH 7) within 50 cm

h Acid 10-60 % Al saturation of CEC by (" bases and unbuffered Al) within 50 cm or pH 1:1 H_2O between 5.0 to 6.0

i Fe-p fixation% free Fe_2O_3/ % clay > 0.2 or hues redder than 5 YR and granular structure

x X-ray amorphouspH > 10 in 1N NaF or positive to field NaF test or other indirect evidences of allophane dominance in clay fraction

v VertisolVery sticky plastic clay >35 % clay and > 50 % of 2:1 expanding clays; COLE > 0.09. severe topsoil shrinking and swelling.

k K deficient<10 % weatherable minerals in silt and sand fraction within 50 cm or exch. K < 0.20 meq/100 g or K < 2 % of " of bases, if " of bases < 10 meq/100

b Basic ReactionFree $CaCo_3$ within 50 cm or pH > 7.3.

s Salinity4 mmhos/cm of saturated extract at 25 C within 1 metre.

n Natric > 15 % Ma satiratopm pf CEC within 50 cm

c Cat claypH is 1:1 H_20 is < 3.5 after drying. Jarosite mottles with huse 2.5Y or yellower and chromas 6 or more within 60 cm.

g Gravel ; a prime (') denotes 15-35 % gravel or coarser (>2mm) particles by volume to any type or substrata (example S'L= gravelly, sand over loamy; SL' = sandy over gravelly loam); two prime marks (") denote more than 35% gravel of coarser particles (>2mm) by volume in any type or substrata type (example LC"= Loamy over clayey skeletal; L'C" = gravelly loam over clayey skeletal)

% Slope, where it is desirable to show slope with the FCC, the slope range percentage can be placed in parentheses after the last condition modifier (example: Sb(0-8%)= uniformly sandy soil calcareous in reaction, 0-8 %slope)

4.4. Land Irrigability Classification

In this, soils are classified according to their suitability for sustained use under irrigation based on soil properties. This system can be used for selection of irrigable lands, estimation of water requirements, development costs and benefits from irrigation. In LIC special attention is paid to three factors namely

1. Quality and quantity of water
2. Drainability of land (Drainage requirement), and
3. The economic consideration (cost of inputs for land development).

1. Quantity of water

a) Equilibrium salinity level

b) Equilibrium exchangeable sodium percentage

c) Availability of water to the land relation to water requirement.

2. Drainage requirement

a) Permeability of substrata and feasibility of providing needed drainage.

b) Cost of drainage measures.

3. Other economic considerations

a) Production cost and yield potential.

b) Land development costs.

c) Other factore affecting benefit cost ratio.

Apart from the above factor the other suitability criteria were also considered to classify the lands are:

- Available water holding capacity (AWHC)
- Effective rooting depth
- Intake characteristics of soils.

LIC is the interpretation at land and soil characters for potential irrigation. It enables to demarcate the area suitable for irrigation. It has three categories viz.,

i) Land irrigability class (LIC)

ii) Land irrigability sub class (LISC)

iii) Land irrigability units

Land Irrigability Classes

Class 1 Lands that have few limitations for sustained use under irrigation. The soils in this classs are nearly having deeper rooting zones, have favaourable permeability, texture, and available moisture holding capacity.

Class 2 Lands that has moderate limitations for sustained use under irrigation. Limitations may be single or combination of factor like 1.very gently, 2. Less than ideal depth, texture permeability, 3. Moderate salinity or alkalinity, 4. Some what unfavourable topography of drainage conditions.

Class 3 Lands that have severe limitations for sustained use under irrigation. Limitations are, 1. Gently slopes, 2. Unfavourable soil depth, texture, permeability, 3. Moderately server salinity or alkalinity, 4. Unfavourable topography

Class 4 Lands that is marginal for sustained use under irrigation because of very severe limitation. Limitations are 1. Moderately steep slopes, 2. Very unfavourable soil depth, texture, permeability 3. Severe salinity/ alkalinity when in equilibrium with irrigation water, 4. Very unfavourabletopography.

Class 5 Lands that are temporarily classified as not suitable for sustained use under irrigation pending further investigation.

Class 6 Lands not suitable for sustained use under irrigation.

LIC are worked out based on soil irrigability class (SIC), topography and drainage. There are five soil irrigability classes indicated by letters A to E

SIC – A → None to slight soil limitation for sustained irrigation

SIC – B → Moderate soil limitation for sustained irrigation

SIC – C → Severe soil limitation for sustained irrigation

SIC – D → Very severe limitation for sustained irrigation

SIC – E → Unsuitable for irrigation

Land irrigability sub classes

These groups of land irrigability units that have same kinds of dominant limitations for sustained use under irrigation when lands are placed in aclass lower than1, the reasons should be indicated whether limitation is due to soil (s), topography (t), drainage (d). The lands with one or more limitations may also be indicated with relevant letters after class. Eg. Class 2 st.

It means land irrigability class 2 with the limitations are soil and topography.

Land irrigability units

Land irrigability units are grouping of lands that are nearly alike in suitability for irrigation and having similar crop adaption, yield potential and soil and water management needs. Grouping of mapping units with soil irrigability class indicated in the land irrigability class. (soil irrigability class indicated in brackets which indicating LIC)

Eg. Land irrigability class of a soil is 2 s(b)

2 – Land irrigability class

S – Land irrigability sub class

(b) – Soil irrigability class

4.5. Storie Index

Earl Storie (1954) proposed Storie Index for evaluation of soil. Storie index of soil rating is a function of 4 major factors. They are

A - Rating on characteristics of physical profile condition.

B - Rating on the basis of surface texture

C - Rating on the basis of slope.

X - Rating of conditions not covered under A, B, C. Such as drainage, fertility, erosion, acidity / alkalinity etc.

Rating = A x B x C x X

It is a method of soil rating based on soil characteristics that govern the land's potential utilization and productivity capacity. It is independent of other physical or economic factors that might determine the desirability of growing certain plants in a given location. The evaluation is easy to be realized and a variety of categories are comprised in few categories.

i) Development stage I – soils on recent alluvial having A – C profile

ii) Development stage II – soils on young alluvial having A –(B)- C profile

iii) Development stage III – soils on older alluvial having A –Bt- C profile

iv) Development stage IV – soils on older plains or terraces having strongly developed profiles

v) Development stage V – soils on older plains having hard pan subsoil

vi) Development stage VI – soils on older terraces and upland with clayey subsoils

vii) Development stage VII – soils on upland underlain by hard igneous bed rocks

viii) Development stage VIII – soils on upland underlain by hard sediementary rocks

ix) Development stage IX – soils on upland underlain by softly consolidated rocks

Storie Index rating = [A/100 x B/100 x C/100 x X/100 x Y/100] x 100

A higher value indicates high productive soil. In each of A, B, C and X based on properties required for optimum growth of crops, different groupings done and accordingly the marks are awarded. Here, since the rating is working out on multiplication pattern; the limiting factor will decide the index rating to a higher magnitude which is highly useful phenomena in working out of the soil index rating.

4.6. Productivity Rating

Present productivity rating

This is a parametric system of assessing the land productivity of soils based on soil and site parameters that influence the yields through mathematical equations. The parametric model developed by Riquier *et al.* (1970) has been employed for assessment of soil productivity where in nine soil/site factors were used in computing soil Productivity Index (PI).

PI = H x D x P x T x N x S x O x A x M

where, H - soil moisture content, D - drainage, P - effective soil depth, T - texture / structure, N - base saturation, S - soluble salts, O - organic matter and A - mineral exchange capacity and M- mineral reserves.

Mineral exchange capacity and nature of the clay in the B Horizon (A)

The soil clay mineralogy plays a critical role in cation exchange capacity and indicative of the fertility level of the soils.

$$\text{Exchange capacity of clay} = \frac{(\text{CEC of soil-K X \%OM}) \text{ X } 100}{\%\text{Clay}}$$

Where

CEC= Cation exchange capacity (at a depth of 60 cm for crops and forest/other trees and 30 cm for pasture)

K = constant factor, 1.5 for tropical soils

OM= Organic matter content

Each factor is rated on a scale from 0 to 100, the actual percentages are being multiplied by each other. The resultant index of productivity is set against a scale placing the soils in one of the five productivity classes.

Productivity class	Marks	Interpretation
I	65 – 100	Excellent
II	35 – 64	Good
III	20 – 34	Average
IV	8 – 19	Poor
V	0 – 7	Extremely poor

Potential productivity rating (P')

After effecting all the possible soil improvement factors, the potential productivity rating was worked out and grades were assigned (Riquier et al., 1970).

Co-efficient of improvement

The ratio of P : P' indicating the extent to which productivity can be improved is called the co-efficient of improvement. The co-efficient of improvement (CI) was worked out based on the present productivity and potential productivity ratings as given below:

$$\text{Co-efficient of Improvement (CI)} = \frac{\text{Potential productivity rating}}{\text{Present Productivity rating}}$$

If the ratio is 2.5 means, the productivity can be improved by 2.5 times more.

Here this is indicated in different colour while preparing the productivity maps. Eg. Light grey indicates Class I, Yellow indicates Class II.

4.7. Land Suitability Classification for Field Crops

Each plant species require definite soil and site conditions for its optimum growth. Although some plants may be found to grow under different soils and extreme agro-ecological conditions, all the plants cannot grow in the same soil and under the same environment. The success or failure of a plant species to grow in a particular soil and site is determined largely by the climate, availability of water and plant nutrients. Water and nutrient availability are largely controlled by the soil physical, chemical and physico-chemical properties. Based on the soil properties and climatic data, the suitability of the land to support different crops can be worked out following the FAO guide lines (1976).

A. Suitability criteria

Most of the plant species need well drained, moderately fine to medium textured soils, free of salinity and having optimum physical environment. Soil resource maps help in predicting the behavior and suitability of soils for growing field crops and forest or plantation crops, once the suitability criteria is established. The suitability of a land to support a crop is classified into different categories consisting of Order, Class, Subclass and Unit.

Order

Land Suitability Orders denote the kind of suitability. There are two land suitability orders namely 'S' and 'N'. 'S' denotes that the land is suitable for crop cultivation and 'N' denotes unsuitable.

Class

The classes denote the degree of limitations. There are 3 classes under the order S (S1, S2 and S3) and 2 classes under the order N (N1 and N2) reflecting the degree of suitability within the order.

Class 1	S1	Highly suitable (Green)	Land having no significant limitations to sustained application of a given use, or only minor limitations that will not significantly reduce productivity of benefits and will not raise inputs above an acceptable level
Class2	S2	moderately suitable (Yellow)	Land having limitations which in aggregate are moderately severe for sustained application of a given use; the limitations will reduce productivity of benefits and increase required inputs to the extent that the overall advantage to the gained from the use, although still attractive, will be appreciably inferior to that expected on class S1 land
Class 3	S3	Marginally suitable (Brown)	Land having limitations which in aggregate are severe for sustained application of a given use and will so reduce productivity or benefits or increase required inputs that this expenditure will be only marginally justified.
Class1	N1	Currently not suitable (Red)	Land having limitations which may be surmountable in time but which cannot be corrected with existing knowledge at currently acceptable cost; the limitations are so severe as to preclude successful sustained use of the land in the given manner

Class 2	N2	Permanently not suitable (Violet)	Land having limitations which appear so severe as to preclude any possibilities of successful sustained use of the land in the given manner

Sub class

The sub-classes reflect the kind of limitations or the improvement measures required within a class. These are indicated by symbols using lower case letters following the arabic numerals used for classes. The following are the sub-classes:

1. Climate (c) (included different weather parameters)
2. Topography and landscape (t)
3. Wetness conditions of soil (w)
   - Drainage
   - Flooding
4. Physical conditions of soil (s)
   - Texture
   - Gravel / stoniness (Surface soil, subsurface soils)
   - Depth
   - Lime / Calcium carbonate
   - Gypsum
5. Soil fertility (f) (not easily correctable)
   - Organic matter
   - Cation exchange capacity
   - Base saturation
   - Nutrient availability
6. Salinity and alkalinity (n)
   - Salinity
   - Groundwater quality
   - Alkalinity / Sodicity

Unit

The land suitability unit is similar in management requirements. This suggests the relative importance of land improvement works. It is indicated by Arabic numerals enclosed in paranthesis, following the sub class symbol.

Example

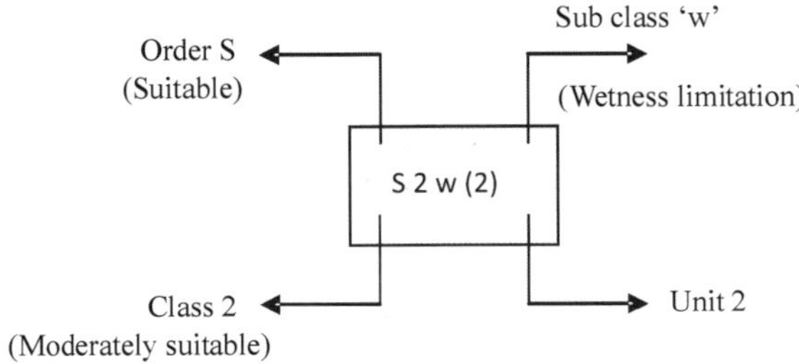

Capability index

Sys and Verhese (1975) proposed the following Capability Index (Ci) based on nine parameters for crop production in the arid and semi-arid regions.

Ci = ABCDEFGHI

Where

A - Soil texture

B - Calcium carbonate

C - Gypsum

D - Salinity

E - Sodium saturation

F - Drainage

G - Soil depth

H - Epipedon

I - Weathering stage / profile development

Evaluation based on degree of limitation

No limitation (0)	:	Characteristics are optimal for plant growth (Ci$\geq$ 80)
Slight limitation (1)	:	Characteristics are nearly optimal for the land utilization type and adversely affect productivity by not more than 20 per cent from the optimal yield (Ci : 60 to 80)

Moderate limitation (2) : Characteristics that have moderate influence on crop yield decline, but benefits can still be made and the land use remains economical (Ci : 45 to 60)

Severe limitation (3) : Characteristics that have such an influence on crop yield decline that use becomes uneconomical for the considered land utilization type (Ci : 30 o 45)

Very Severe limitation (4) : Characteristics that decreases the yield below profitable level and inhibits the use of soil for the considered land utilization (Ci : <30)

5

Land Use Planning

Land is merely an area of the earth's surface comprising all physical, chemical and biological environments that influence the land use (FAO ,1976). Land refers not only to soil but also to climate, landforms, hydrology, geology, natural vegetation including fauna and flora. The land use simply means management of a land for specific uses to meet the human needs in all rural, urban and industrial context, whereas the land use plan refers to a coherent set of decisions about the ways to achieve the desired level preferably on economic background. Thus, the land use planning is the systematic evaluation and assessment of a land and associated attributes for the purpose of identifying the best land use options beneficial to landowners or users without degrading the land resources or environment. It includes not only the landowner or users, but also the policy or decision makers as well as economic experts for financial measures from planning stage or logistic planning to execution.

Land cover

Relates to the type of feature present on the surface of the earth e.g., Fields, trees, concrete high ways.

Land use

Relates to the human activity associated with a specific piece of land. The same land would have a land cover consisting of roofs, pavements, grass and trees. The land use in an area is really an intervention by the people over the existing natural eco-environment for satisfying their needs.

Land cover is distinct from land use despite the two terms often being used interchangeably. Land use is a description of how people utilize the land and socio-economic activity - urban and agricultural land uses are two of the most commonly recognised high-level classes of use. At any one point or place, there may be multiple and alternate land uses, the specification of which may have a political dimension.

Land Use Planning

Land-use planning is the systematic assessment of land and water potential, alternatives for land use and economic and social conditions in order to select and adopt the best land-use options. Its purpose is to select and put into practice those land uses that will best meet the needs of the people while safeguarding resources for the future. The driving force in planning is the need for change, the need for improved management or the need for a quite different pattern of land use dictated by changing circumstances.

When is land-use planning useful?

Two conditions must be met if planning is to be useful:

- The need for changes in land use, or action to prevent some unwanted change, must be accepted by the people involved;
- There must be the political will and ability to put the plan into effect.

Where these conditions are not met, and yet problems are pressing, it may be appropriate to mount an awareness campaign or set up demonstration areas with the aim of creating the conditions necessary for effective planning.

5.1 Objectives of Land Use Planning

The land use planning goals define what is meant by the 'best' use of the land and may be grouped under the headings of efficiency (land use must be economically viable while making effective and productive use of the same), equity and acceptabaility (land use must be socially acceptable) and sustainability (simultaneous combination of production and conservation.

The objectives of land use planning can be summarized as follows:

- To quantify the agro-ecology factors that define the categories of land use and optimal patterns.
- To integrate the economic/market relationships of the input-output matrix that governs the acreage and production relations under the particular crop regime identified for the category of land use identified.
- As a tool for bench-marking farmers' managerial index and their capacity to absorb capital and technology and charting out ways and means of enhancing it.
- To identify the institutional infrastructure needs for promoting 'brand equity' and usher in the commercial/industrial status to agriculture.

- To help in reorienting the departmental perspective from present top-to-bottom departmental approach to a bottom-up management process with the participation of all stake holders in agricultural development.

Making the best use of limited resources

Our basic needs of food, water, fuel, clothing and shelter must be met from the land, which is in limited supply. As population and aspirations increase, so land becomes an increasingly scarce resource.

Land must change to meet new demands yet change brings new conflicts between competing uses of the land and between the interests of individual land users and the common good. Land taken for towns and industry is no longer available for farming; likewise, the development of new farmland competes with forestry, water supplies and wildlife.

Planning to make the best use of land is not a new idea. Over the years, farmers have made plans season after season, deciding what to grow and where to grow it. Their decisions have been made according to their own needs, their knowledge of the land and the technology, labour and capital available. As the size of the area, the number of people involved and the complexity of the problems increase, so does the need for information and rigorous methods of analysis and planning.

However, land-use planning is not just farm planning on a different scale; it has a further dimension, namely the interest of the whole community.

Land-use planning aims to make the best use of limited resources by:

- Assessing present and future needs and systematically evaluating the land's ability to supply them;
- Identifying and resolving conflicts between competing uses, between the needs of individuals and those of the community, and between the needs of the present generation and those of future generations;
- Seeking sustainable options and choosing those that best meet identified needs;
- Planning to bring about desired changes;
- Learning from experience.

There can be no blueprint for change. The whole process of planning is iterative and continuous. At every stage, as better information is obtained, a plan may have to be changed to take account of it.

5.2 Planning at Different Levels

Land-use planning can be applied at three broad levels: national, district and local. These are not necessarily sequential but correspond to the levels of government at which decisions about land use are taken.

Different kinds of decision are taken at each level, where the methods of planning and kinds of plan also differ. However, at each level there is need for a land-use strategy, policies that indicate planning priorities, projects that tackle these priorities and operational planning to get the work done.

The greater the interaction between the three levels of planning, the better. The flow of information should be in both directions (Fig. 5.1). At each successive level of planning, the degree of detail needed increases, and so too should the direct participation of the local people.

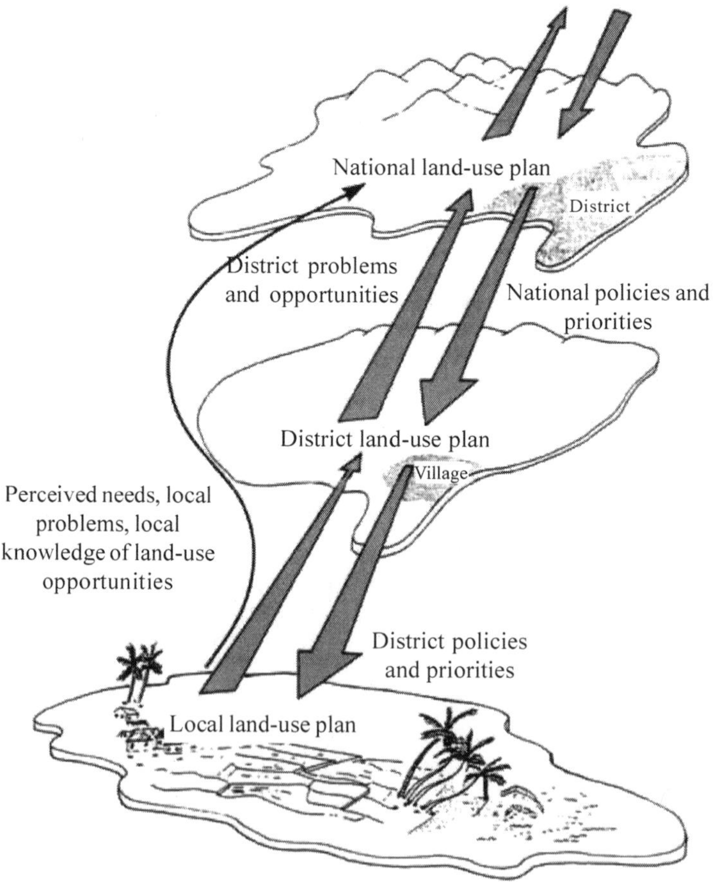

Fig. 5.1 Two-way links between planning at different levels

National level

At the national level, planning is concerned with national goals and the allocation of resources. In many cases, national land-use planning does not involve the actual allocation of land for different uses, but the establishment of priorities for district-level projects. A national land-use plan may cover:

- Land-use policy: balancing the competing demands for land among different sectors of the economy food production, export crops, tourism, wildlife conservation, housing and public amenities, roads, industry;
- National development plans and budget: project identification and the allocation of resources for development;
- Coordination of sectoral agencies involved in land use;
- Legislation on such subjects as land tenure, forest clearance and water rights.

National goals are complex while policy decisions, legislation and fiscal measures affect many people and wide areas. Decision-makers cannot possibly be specialists in all facets of land use, so the planners' responsibility is to present the relevant information in terms that the decision-makers can both comprehend and act on.

District level

District level refers not necessarily to administrative districts but also to land areas that fall between national and local levels. Development projects are often at this level, where planning first comes to grips with the diversity of the land and its suitability to meet project goals. When planning is initiated nationally, national priorities have to be translated into local plans. Conflicts between national and local interests will have to be resolved. The kinds of issues tackled at this stage include:

- The siting of developments such as new settlements, forest plantations and irrigation schemes;
- The need for improved infrastructure such as water supply, roads and marketing facilities;
- The development of management guidelines for improved kinds of land use on each type of land.

Local level

The local planning unit may be the village, a group of villages or a small water catchment. At this level, it is easiest to fit the plan to the people, making use of local people's knowledge and contributions. Where planning is initiated at the district level, the programme of work to implement changes in land use or management has to be carried out locally. Alternatively, this may be the first level of planning, with its priorities drawn up by the local people. Local-level planning is about getting things done on particular areas of land - what shall be done where and when, and who will be responsible.

Starting at the local level: bottom-up planning

"Bottom-up" planning is initiated at the local level and involves active participation by the local community. The experience and local knowledge of the land users and local technical staff are mobilized to identify development priorities and to draw up and implement plans.

The advantages are

- Local targets, local management and local benefits. People will be more enthusiastic about a plan seen as their own, and they will be more willing to participate in its implementation and monitoring;
- More popular awareness of land-use problems and opportunities;
- Plans can pay close attention to local constraints, whether these are related to natural resources or socio-economic problems;
- Better information is fed upwards for higher levels of planning

The disadvantages are that

- Local interests are not always the same as regional or national interests;
- Difficulties occur in integrating local plans within a wider framework;
- Limited technical knowledge at the local level means technical agencies need to make a big investment in time and labour in widely scattered places;
- Local efforts may collapse because of a lack of higher-level support or even obstruction.

Planning at these different levels needs information at different scales and levels of generalization. Much of this information may be found on maps. The most suitable map scale for national planning is one by which the whole country fits on to one map sheet, which may call for a scale from 1:5 million to 1:1 million or larger. District planning requires details to be mapped at about 1:50000, although some information may be summarized at smaller scales, down to 1:250000.

For local planning, maps of between 1:20000 and 1:5000 are best. Reproductions of air photographs can be used as base maps at the local level, since field workers and experience show that local people can recognize where they are on the photos.

5.3 Steps in Land-use Planning

Every land-use planning project is different. Objectives and local circumstances are extremely varied, so each plan will require a different treatment. However, a sequence of ten steps has been found useful as a guide. Each step represents a specific activity, or set of activities, and their outputs provide information for subsequent steps. Sequence of ten steps has been found to be useful as a guide for attempting LUP by the Food and Agriculture Organization of the FAO (FAO 1993). The steps are:

Step 1 ***Establish goals and terms of reference:*** Ascertain the present situation; find out the needs of the people and of the government; decide on the land area to be covered; agree on the broad goals and specific objectives of the plan; settle the terms of reference for the plan.

Step 2 ***Organize the work:*** Decide what needs to be done; identify the activities needed and select the planning team; draw up a schedule of activities and outputs; ensure that everyone who may be affected by the plan, or will contribute to it, is consulted.

Step 3 ***Analyse the problems:*** Study the existing land-use situation, including in the field; talk to the land users and find out their needs and views; identify the problems and analyse their causes; identify constraints to change.

Step 4 ***Identify opportunities for charge:*** Identify and draft a design for a range of land-use types that might achieve the goals of the plan; present these options for public discussion.

Step 5 ***Evaluate land suitability:*** For each promising land-use type, establish the land requirements and match these with the properties of the land to establish physical land suitability.

Step 6 ***Appraise the alternatives: Environmental, economic and social analysis.*** For each physically suitable combination of land use and land, assess the environmental, economic and social impacts, for the land users and for the community as a whole. List the consequences, favourable and unfavourable, of alternative courses of action.

Step 7 ***Choose the best option:*** Hold public and executive discussions of the viable options and their consequences. Based on these discussions and the above appraisal, decide which changes in land use should be made or worked towards.

Step 8 ***Prepare the land-use plan:*** Make allocations or recommendations of the selected land uses for the chosen areas of land; make plans for appropriate land management; plan how the selected improvements are to be brought about and how the plan is to be put into practice; draw up policy guidelines, prepare a budget and draft any necessary legislation; involve decision-makers, sectoral agencies and land users.

Step 9 **Implement the plan:** Either directly within the planning process or, more likely, as a separate development project, put the plan into action; the planning team should work in conjunction with the implementing agencies.

Step 10 ***Monitor and revise the plan:*** Monitor the progress of the plan towards its goals; modify or revise the plan in the light of experience.

In a still broader view, the steps can be grouped into the following logical sequence:

- Identify the problems. *Steps 1-3.*
- Determine what alternative solutions exist. *Steps 4-6.*
- Decide which is the best alternative and prepare the plan. *Steps 7-8.*
- Put the plan into action, see how it works and learn from this experience. *Steps 9-10.*

The land use planning cycle is schematically presented in fig. 5.2.

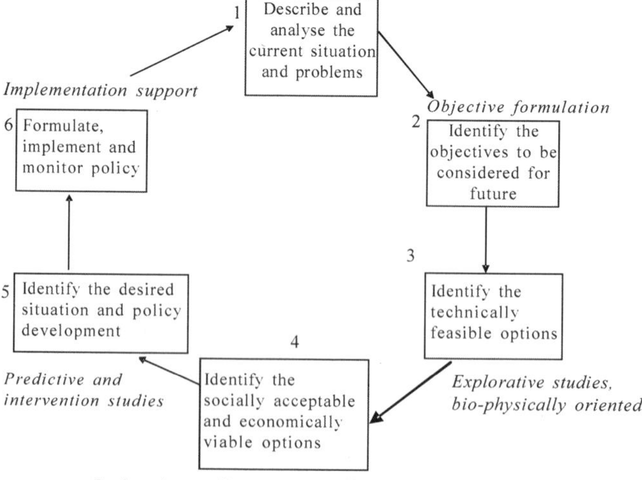

Fig. 5.2 Land use planning cycle

5.4 Land Use Planning Sycle (*Source*: Van Keulen et al. 2000)

The basic components of LUP are biophysical and socioeconomic parameters which are collected through standard survey. Besides the crop husbandry, LUP should also include in its planning process all the allied subjects like horticulture, agroforestry, animal husbandry and livestock, fishery poultry etc. in an integrated way for the benefit of farmers and strengthening agro-ecology.

Land use planning is a scale-dependent process and therefore there is a need for planning at different levels viz, National, District, Local needs information at different scales and levels of generalization. Much of this information may be found on maps. The most suitable map scale for national planning is one by which the whole country fits on to one map sheet, which may call for a scale from 1.5 million to 1:1 million or larger. District planning requires details to be mapped at about 1:50,000; although some information may be summarized at smaller scales, down to 1:250,000. for local planning, maps of between 1:10000 and 1:5000 are best.

Table 5.1 Planning levels and recommended map scales.

Level	Administrative unit	Map scale
National	Country	Small: 1: 250,000
		Medium : 1: 1000 000
		Large: 1: 5000 000
Sub-national (meso)	Region, Province, district	Small: 1: 100,000
		Medium : 1: 250 000
		Large: 1: 1 000 000
Local	Sub-district, village, community	Small: 1: 10,000
		Medium : 1: 25 000
		Large: 1: 50 000
Farm	Farm	Small: 1: 1 000
		Medium : 1: 5 000
		Large: 1: 10 000

In the context of regional agricultural development, land use planning is the sectoral allocation of land to optimize the postulated objectives under the existing environmental and social opportunities and constraints. At regional level resource managers (development planners, decision and policy makers) are often concerned about what new agricultural (including livestock) production innovation/ activities need to be promoted/supported under varying constraints of land (including quality), labour, water, capital and the adverse effect of agricultural/ land use (economic) activities on the quality of natural resources.

For undertaking these complex tasks they need the exploratory land use analysis and decision support system (DSS) which use a rational, scientific analysis and evaluation of different land use options. This decision support system must evaluate land resources, socio-economic conditions of land users, current and future lifestyle and policy options for land use planning. Development of such a decision support system for land use planning in the context of multifunctional agriculture would involve the following steps:

- To delineate rainfed, arid, coastal, irrigated and hill and mountain agro-ecosystems.
- Interaction and workshops with stake holders- state government officials, NGOs, KVK officials, other development agencies etc. to identify agricultural objectives, constraints and priorities under different present and future scenarios, collection and characterization of biophysical and socio-economic resources.
- Evaluation of the land suitability for each production activity (crop, crop+ livestock/poultry/ fisheries etc.)
- Fixing the demands for agricultural products and their targets using agricultural production resources (land, water etc.) based on diagnostic survey, policy documents, action plans, trend for growth analysis, know-how by experts and subject matter specialists such as agriculturists, horticulturists, geographers, environmentalists, economists, sociologiests and management experts.
- Yield estimation and yield gap analysis for various land use options
- Interactive multiple goal linear programming (IMGLP) to enable policy makers to analyse land use options and their consequences.

5.5 Limitations of Current Land Use Planning Procedure

- Applying top-down approach
- Socioe-economic conditions are not considered.
- Participatory approach is not adopted to involve all stake holders
- Technology element is not considered
- Relations in process of land use are not quantified
- Multiple scales are not integrated in land use plans
- Multidisciplinary work needs to be encouraged
- Feasibility of a plan is low

5.6 Holistic Sustainable Land Management Model Towards Land Use Planning

The conceptual model for holistic land management is given in Fig. 5.3

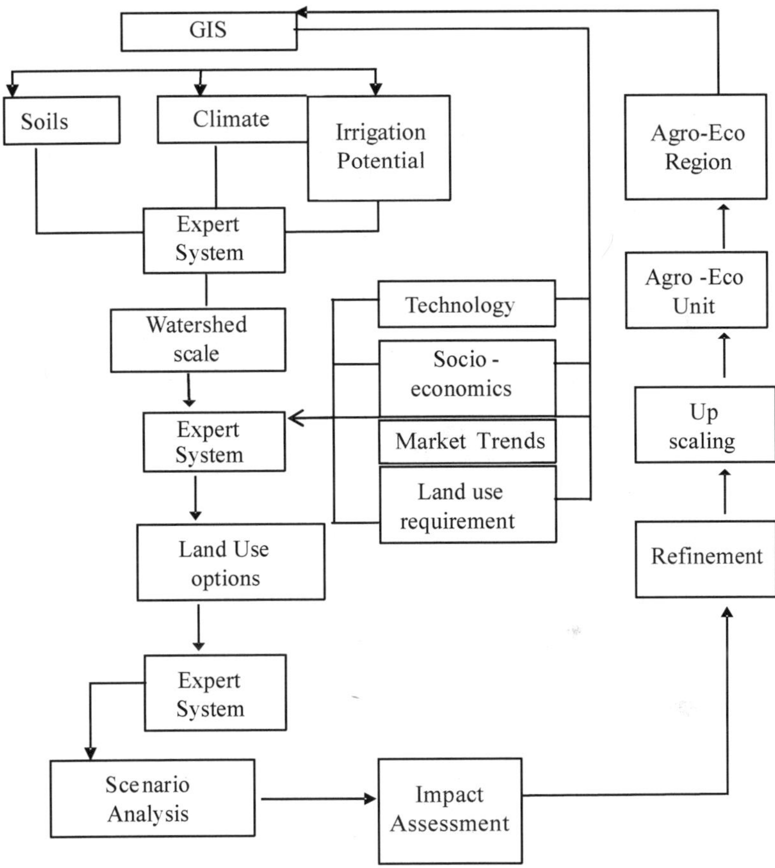

Fig. 5.3 Conceptual model for holistic sustainable land management

Step wise procedure for the execution of the model is described hereunder..

Step 1 : Identify and analyse the extent and severity of constraints of soil and water resources. Careful analyse of climatic parameter, available technologies to avert the constraints, socio-economic conditions of farmers and their payment capacity, people demand and market trend is required.

Step 2 : Rule based integration of soil and water resources with increasing level of constraints on watershed scale with the emphasis on water flow, flow direction, residence time, moisture surplus, storage and period of moisture availability.

Step 3 : Physical suitability of identified units for different land use, using threshold limit of different attributes of land. The stress should be focused on water related properties that govern the performance of any kind of land use under set of conditions.

Step 4 : Climatic conditions and crop properties are used in the crop growth simulation models to calculate yield potentials for the identified units under various land uses with different management techniques in the situation of normal, sub-normal rainfall and drought conditions scenarios in combination to economic, social and political goals with a linear programming model.

Step 5 : Refine and optimize land u se in light of farmer's perception about the different land use scenario; their requirement, present and expected market demand.

Step 6 : Impact assessment and monitoring the performance of suggested land use with remote sensing data

Step 7 : Quantifying the existing database of smaller scale on regional basis with help of pedo-transfer functions generated on landscape scale.

Step 8 : Upscaling the data on their transformation into agro-ecounit to agro-ecological region via zones and sub-zones.